Neodynamics:

or Adaptation and Coherence

in the Universe and Humanity

Benjamin James

November 2024

Table of Contents

Introduction
Purpose and Scope

The foundational purpose of Neodynamics is to formalize a framework for adaptive systems capable of thriving in complex, interconnected environments. Traditional systems design, which often prioritizes stability or optimization under fixed conditions, struggles to handle the dynamic and unpredictable changes characteristic of modern challenges. Neodynamics addresses this gap by focusing on adaptability, emergent coherence, and feedback-driven processes as essential principles for system resilience and innovation.

The scope of Neodynamics spans both theoretical foundations and practical applications. Drawing from interdisciplinary fields such as systems theory (Bertalanffy, 1968), thermodynamics (Prigogine, 1980), and information theory (Shannon, 1948), it synthesizes these ideas into a cohesive paradigm. Its two primary constructs, the Unified Field of Adaptive Potential (UFAP) and the SPARC Framework, provide tools to model, design, and understand adaptive systems. These constructs enable systems to explore potential responses dynamically and refine decisions iteratively, ensuring alignment with shifting objectives while maintaining coherence.

The need for such a framework is underscored by current global challenges. Static systems—technological, organizational, or ecological—often fail when confronted with rapid technological advancements, climate crises, and geopolitical instability. For example, governance systems frequently exhibit rigidity during crises like pandemics, resulting in inefficient responses and cascading failures (Ostrom, 2009). Similarly, disruptions in healthcare and supply chains demonstrate the limitations of conventional models in managing unpredictability (Ivanov & Dolgui, 2020). Neodynamics offers a pathway to designing systems that are not only resilient but also capable of continuous recalibration and learning.

This text serves three key objectives. First, it articulates the theoretical foundations of Neodynamics by integrating principles from multiple disciplines, offering a cohesive conceptual framework. Second, it introduces the UFAP and SPARC Framework as tools for designing systems capable of dynamic adaptation. Third, it explores speculative applications across diverse domains, from artificial intelligence to governance and healthcare, demonstrating the framework's transformative potential.

Unlike traditional systems approaches, which often treat complexity as a problem to be minimized, Neodynamics views complexity as an opportunity. It emphasizes that emergent coherence—the natural alignment of a system's components to achieve a functional whole—can arise through diversity, feedback loops, and recursive decision-making processes. This perspective aligns with modern understandings of complex adaptive systems (Holland, 1992) and provides a foundation for innovation across multiple fields.

Neodynamics is positioned as a unifying paradigm for adaptive systems, bridging theoretical research and practical implementation. By addressing the limitations of static models and embracing the dynamics of complex environments, it lays the groundwork for systems that thrive in an era of constant change.

The Problem of Static Systems

Static systems are defined by their reliance on fixed architectures, predetermined processes, and narrow optimization goals. While this rigidity provides stability in controlled and predictable environments, it becomes a liability in dynamic, interconnected, and uncertain contexts. Static systems fail not only because they are slow to adapt but also because their foundational designs resist the very mechanisms—such as feedback integration and dynamic recalibration—that are essential for survival in complex environments. These failures are increasingly evident across technological, ecological, and social systems.

At the heart of this failure lies the inability of static systems to manage complexity. Complexity, in the context of interconnected systems, arises from interactions between diverse components that are nonlinear, emergent, and often unpredictable (Bar-Yam, 1997). Traditional approaches to system design treat complexity as an obstacle to be simplified or controlled. For instance, linear models in governance, economics, and engineering often ignore emergent properties, focusing instead on optimizing specific variables within rigid constraints. This reductionist mindset produces brittle systems that are efficient in narrow contexts but incapable of adapting to disruptions or novel scenarios.

Governance systems exemplify this rigidity. Most governments operate within hierarchies that prioritize long-term stability and efficiency over adaptability. Such systems perform poorly under crisis conditions, where rapid recalibration is critical. During the COVID-19 pandemic, governments adhering to static policy frameworks struggled to respond effectively to shifting circumstances, such as changing epidemiological data and public behavior. The delayed implementation of adaptive measures—like dynamic lockdowns or real-time vaccine distribution adjustments—led to cascading failures in public health and trust (Ostrom, 2009). These failures reflect a broader inability of static governance systems to incorporate feedback, explore alternative pathways, and recalibrate objectives dynamically.

In the technological domain, artificial intelligence systems further illustrate the pitfalls of static design. AI models, particularly those designed with fixed optimization goals, often fail to generalize beyond their training data. This limitation arises because static AI systems are bound by narrow assumptions about their operational environments. For example, diagnostic AI in healthcare has struggled with changing datasets or patient populations, leading to biases and erroneous predictions (Barocas et al., 2019). Static design, rooted in rigid architectures, precludes the possibility of self-correction through feedback, leaving these systems vulnerable to shifts in context that were not anticipated during their development.

In ecological systems, the consequences of static frameworks are equally profound. Climate change, for instance, has exposed the vulnerability of agricultural and resource management systems designed around historical weather patterns. Static irrigation and crop selection models, optimized for stable climates, now face catastrophic failures as weather extremes become more frequent and unpredictable (Rockström et al., 2009). The rigidity of these systems prevents adaptation, resulting in diminished yields, resource scarcity, and environmental degradation. By contrast, ecosystems themselves often exhibit adaptive resilience through feedback-driven processes, such as predator-prey dynamics, which enable them to recalibrate and stabilize under changing conditions.

Supply chains, structured around static optimization principles like just-in-time inventory, offer a final and stark example. While these models maximize efficiency under stable conditions, they fail catastrophically under disruption. The COVID-19 pandemic disrupted global supply chains by introducing rapid demand shocks and supply bottlenecks. Static models, incapable of integrating real-time feedback, resulted in widespread shortages, inefficiencies, and economic losses (Ivanov & Dolgui, 2020). These failures demonstrate the broader fragility of systems that prioritize narrow optimization over adaptability.

In each of these examples, static systems demonstrate an inability to explore alternative pathways, integrate dynamic feedback, or recalibrate objectives when faced with uncertainty. Neodynamics offers a response to these limitations, proposing a shift from rigidity to dynamic adaptability. By embedding principles of emergent coherence, recursive choice, and feedback-driven recalibration, Neodynamics seeks to design systems capable of thriving under complexity. This represents a fundamental departure from the static paradigms that dominate current system design, offering instead a framework for systems that continuously evolve, learn, and adapt in the face of uncertainty.

Static systems are further constrained by their reliance on predictive control mechanisms. These mechanisms operate on the assumption that future states can be accurately forecasted based on historical data and predefined models. However, in complex, nonlinear systems, predictability diminishes as interactions between components generate emergent behaviors that are not directly tied to initial conditions (Lorenz, 1963). This fundamental limitation, often referred to as the "prediction horizon," underscores the inadequacy of static systems in managing environments characterized by uncertainty and rapid change.

One illustrative example of this limitation is seen in financial systems. Traditional risk models, such as those used in banking and investment management, often rely on static assumptions about market behavior. The 2008 global financial crisis exposed how rigid optimization frameworks, such as value-at-risk models, failed to account for the systemic risks created by interconnected markets and cascading failures. Static systems in finance amplified rather than mitigated the crisis, as their inability to recalibrate in real-time exacerbated liquidity shortages and market instability (Haldane & May, 2011). This demonstrates how static design not only fails to adapt but can actively propagate failure across interconnected networks.

The natural sciences provide a contrasting paradigm in which adaptability is central to system survival. Biological ecosystems, for instance, operate through distributed processes that dynamically balance competition and cooperation among organisms. This adaptability enables ecosystems to maintain functional coherence despite disturbances such as habitat loss or species extinction. For example, the phenomenon of ecological succession demonstrates how ecosystems reorganize themselves following a disruption, gradually stabilizing into new equilibria through feedback loops and emergent behaviors (Odum, 1969). Static human systems lack these mechanisms of self-organization, rendering them incapable of achieving similar resilience.

Education systems offer another domain in which static models have proven insufficient. Traditional curricula and pedagogical methods are designed to optimize learning outcomes for standardized student populations, yet they fail to address the diverse and evolving needs of individual learners. Rigid educational frameworks, driven by standardized testing and fixed content delivery, often result in disengagement and inequitable outcomes. Adaptive learning technologies, in contrast, have begun to demonstrate how feedback-driven processes can personalize education, dynamically adjusting to student progress and needs (Heffernan & Koedinger, 2012). However, these technologies remain in their infancy, and their development requires precisely the kind of adaptive frameworks that Neodynamics aims to formalize.

Static systems also suffer from an inherent fragility in their structural design. This fragility arises because such systems optimize for efficiency within narrow performance parameters, leaving them vulnerable to perturbations outside these parameters. Complex adaptive systems, by contrast, prioritize robustness over efficiency, enabling them to absorb shocks and reorganize dynamically. This distinction is crucial in fields such as infrastructure management, where rigid designs often fail under stress. For example, urban water systems optimized for predictable rainfall patterns are increasingly overwhelmed by extreme weather events, resulting in flooding and water shortages (Gersonius et al., 2013). Adaptive infrastructure models, informed by principles like those in Neodynamics, could mitigate these risks by integrating real-time monitoring and dynamic resource allocation.

Ultimately, the failure of static systems lies not only in their inability to adapt but in their fundamental design philosophy, which prioritizes control and predictability over flexibility and evolution. Neodynamics challenges this paradigm by introducing constructs such as the Unified Field of Adaptive Potential (UFAP) and the SPARC Framework, which emphasize exploration, feedback, and emergent coherence. These constructs shift the focus from optimization within fixed conditions to adaptability across dynamic and uncertain environments. In doing so, Neodynamics establishes the groundwork for systems capable of thriving in complexity, marking a necessary evolution in how we approach system design.

Neodynamics in Context

Neodynamics represents the next evolutionary step in systems thinking, emerging as a response to the challenges and limitations faced by earlier paradigms. Its emphasis on adaptability, emergent coherence, and feedback-driven processes reflects a deliberate shift from traditional approaches that often focused on optimizing stability or performance within narrowly defined contexts. To fully understand the significance of Neodynamics, it is essential to situate it within the broader historical and theoretical developments of systems science and its adjacent disciplines.

The roots of systems thinking can be traced to Ludwig von Bertalanffy's General Systems Theory (GST), which sought to unify principles governing the organization of complex systems across disciplines (Bertalanffy, 1968). GST introduced foundational concepts such as interdependence, feedback, and open systems, laying the groundwork for understanding how systems interact with their environments. However, GST largely emphasized equilibrium and homeostasis, assuming that systems strive for a stable state. While valuable in its time, this focus on stability limited GST's applicability to highly dynamic and unpredictable environments. Neodynamics builds upon GST by shifting the focus from static equilibrium to dynamic adaptability, allowing systems to thrive in flux rather than resist it.

In the mid-20th century, cybernetics emerged as a parallel field, offering deeper insights into feedback mechanisms and control in systems (Wiener, 1948). Cybernetics introduced key ideas such as self-regulation, which remains central to modern systems science. However, traditional cybernetics often approached systems from an engineering perspective, prioritizing predictability and control over emergent behavior. This is particularly evident in its applications to automated systems, where feedback loops were designed to maintain predefined objectives rather than explore new possibilities. Neodynamics diverges by integrating emergent coherence as a core principle, recognizing that the most adaptive systems do not merely stabilize but evolve through iterative exploration of their environments.

In more recent decades, complex adaptive systems (CAS) theory has provided a framework for understanding systems composed of multiple interacting agents that collectively exhibit emergent behaviors (Holland, 1992). CAS theory has been particularly influential in fields such as economics, ecology, and social science, where the interdependence of agents produces patterns that cannot be predicted from the behavior of individual components. While Neodynamics shares CAS theory's emphasis on emergence and complexity, it advances the discussion by introducing actionable constructs like the Unified Field of Adaptive Potential (UFAP) and the SPARC Framework. These constructs translate theoretical principles into tools for design and implementation, enabling practitioners to model and engineer adaptive systems rather than merely describe them.

Resilience engineering also informs the context for Neodynamics. Resilience, broadly defined as a system's ability to absorb shocks and recover, has been a key focus in disciplines such

as disaster management and sustainability science (Holling, 1973).
However, resilience often emphasizes bouncing back to a prior state,
neglecting the potential for transformative adaptation. Neodynamics
extends beyond resilience by advocating for systems that not only
recover but also reorganize and evolve in response to disruption,
thereby continuously expanding their adaptive potential.

In this context, Neodynamics emerges as a unifying
paradigm that integrates and advances prior frameworks, addressing
their limitations while building upon their strengths. It provides a
cohesive theoretical foundation for designing systems that are not
merely robust or resilient but dynamically adaptable, ensuring their
relevance in increasingly complex and interconnected environments.

The incorporation of insights from information theory further
situates Neodynamics within the trajectory of systems science.
Claude Shannon's groundbreaking work on quantifying information
provided a mathematical basis for understanding uncertainty and
communication within systems (Shannon, 1948). Information theory
emphasizes the management of uncertainty as a central challenge in
system design, a principle that aligns closely with the goals of
Neodynamics. However, where information theory often focuses on

static transmission efficiency—maximizing the accuracy of communication within given constraints—Neodynamics shifts the emphasis toward dynamic adaptability in the face of unpredictable changes. The Unified Field of Adaptive Potential (UFAP) serves as an extension of this idea, conceptualizing a multidimensional space in which systems can explore potential responses to uncertainty in real time.

Similarly, Neodynamics draws from thermodynamics, particularly the concepts introduced by Ilya Prigogine in his work on dissipative structures. Prigogine's research demonstrated how systems far from equilibrium can self-organize through the dissipation of energy, leading to emergent complexity (Prigogine, 1980). This principle underpins the concept of emergent coherence in Neodynamics, where systems leverage their internal diversity and external interactions to reorganize dynamically in response to changing conditions. Unlike traditional thermodynamic systems that seek equilibrium, Neodynamic systems are designed to operate in a constant state of flux, using feedback and recursive choice to maintain coherence without requiring stasis.

Game theory also plays a pivotal role in contextualizing Neodynamics. The study of strategic interactions among rational agents, game theory has long been used to model competition, cooperation, and decision-making under conditions of uncertainty. Classical game theory, however, often assumes fixed payoff structures and strategies, limiting its applicability to dynamic environments. By contrast, Neodynamics incorporates recursive choice—a feedback-driven process that allows systems to refine their strategies iteratively based on evolving conditions. This dynamic approach aligns with the emerging field of evolutionary game theory, which considers how strategies evolve over time in response to the actions of other agents (Axelrod, 1984).

Beyond these theoretical influences, Neodynamics also addresses gaps in practical system design. Existing frameworks, such as resilience engineering and CAS theory, often describe the properties and behaviors of adaptive systems but fall short in providing actionable tools for their creation. This limitation has resulted in a gap between theory and implementation, where practitioners struggle to translate abstract principles into tangible designs. Neodynamics bridges this divide by offering constructs like

UFAP and SPARC, which are explicitly designed to operationalize adaptability. These constructs enable system designers to model the space of potential actions, evaluate their feasibility and outcomes, and implement feedback loops that continuously recalibrate objectives in response to change.

In its broadest sense, Neodynamics represents a paradigm shift in systems thinking. It redefines the purpose of system design, moving away from optimization within fixed parameters toward dynamic exploration of possibilities. By integrating principles from information theory, thermodynamics, game theory, and other disciplines, Neodynamics provides a comprehensive framework for understanding and engineering systems that are not only resilient but capable of evolving in complex and uncertain environments. As the challenges of the 21st century demand greater adaptability across all domains, Neodynamics positions itself as both a theoretical advancement and a practical necessity.

The novelty of Neodynamics also lies in its rejection of reductionist principles that have historically dominated systems science. Reductionism, which seeks to break down complex phenomena into simpler components for analysis, has proven inadequate for understanding emergent behaviors that arise from the interactions of system elements. This inadequacy is evident in fields such as economics, where models often fail to predict crises because they isolate variables rather than account for systemic interdependence (Farmer & Foley, 2009). Neodynamics, by contrast, adopts a holistic approach, emphasizing the interactions, feedback loops, and recursive processes that allow systems to self-organize and adapt dynamically.

One of the key distinctions of Neodynamics is its emphasis on designability. While earlier paradigms like cybernetics and CAS theory offered descriptive tools to analyze existing systems, Neodynamics provides actionable frameworks for creating adaptive systems from the ground up. The Unified Field of Adaptive Potential (UFAP) allows designers to conceptualize the full range of adaptive actions available to a system, while the SPARC Framework operationalizes this potential through iterative decision-making. These tools not only enable the modeling of adaptive behavior but also guide the practical engineering of systems capable of thriving in uncertain environments.

In the domain of artificial intelligence, for example, Neodynamics challenges the static architectures of traditional machine learning models. Most AI systems are trained on fixed datasets and operate within predefined boundaries, limiting their ability to generalize or adapt when conditions deviate from their training. Generative models in particular struggle to maintain coherence across diverse outputs, often producing irrelevant or inconsistent results when applied to novel tasks (Bengio et al., 2013). Neodynamics addresses these challenges by embedding feedback-driven processes directly into the system architecture, enabling real-time recalibration and iterative refinement. This adaptive capability not only improves performance but also ensures that AI systems remain aligned with evolving user needs and contextual demands.

In governance, Neodynamics offers a framework for creating policies and institutions that evolve dynamically in response to crises and societal shifts. Traditional governance models often prioritize stability and control, resulting in inflexibility during periods of rapid change. Adaptive governance, informed by Neodynamics, emphasizes feedback mechanisms that allow policies to be continuously evaluated and adjusted based on real-time data. For instance, during public health emergencies, a Neodynamic approach could enable governments to deploy resources dynamically, respond to shifting epidemiological trends, and integrate community feedback into decision-making processes. This iterative approach not only enhances responsiveness but also builds trust by demonstrating a capacity for learning and adaptation.

Healthcare is another domain where Neodynamics has transformative potential. Current medical systems often rely on static diagnostic protocols and treatment pathways that fail to account for individual variability or the dynamic progression of disease. Personalized medicine, which seeks to tailor treatments to the unique characteristics of each patient, aligns closely with the principles of Neodynamics. By leveraging constructs like UFAP, healthcare systems can explore a broader spectrum of diagnostic and therapeutic options, continuously refining these choices based on patient outcomes and emerging data. This adaptive approach has the potential to revolutionize medical practice, shifting the focus from reactive treatment to proactive and dynamic care.

In sum, Neodynamics builds upon the theoretical foundations laid by earlier paradigms while addressing their limitations. It integrates insights from multiple disciplines to create a unifying framework that emphasizes adaptability, feedback, and emergent coherence. By bridging the gap between theory and implementation, Neodynamics positions itself as a critical tool for addressing the complexities of the modern world, offering a pathway for systems that are not only resilient but continuously evolving.

The emergence of Neodynamics also reflects a growing recognition that resilience alone is insufficient to address the challenges of modern systems. Traditional resilience frameworks emphasize a system's ability to withstand shocks and recover to its prior state, yet this approach assumes that stability is inherently desirable. In reality, the environments in which systems operate are often transformed by the very shocks they experience, rendering a return to a previous state both impractical and counterproductive (Walker et al., 2004). Neodynamics advances the concept of resilience by incorporating the idea of transformative adaptation: systems not only recover but reorganize to align with newly emerged conditions, expanding their adaptive potential rather than merely preserving their original function.

This distinction is especially critical in ecological systems, where the impacts of climate change necessitate dynamic, adaptive strategies. For instance, ecosystems that undergo disturbances such as deforestation or rising temperatures often reorganize into entirely new configurations, characterized by different species compositions and resource flows. These transitions, while disruptive, are examples of adaptive evolution that enable ecosystems to maintain functionality in altered environments (Holling, 1973). Neodynamics draws inspiration from these natural processes, emphasizing the capacity of systems to explore alternative configurations and self-organize into states of emergent coherence, even under conditions of significant external stress.

Another critical context for Neodynamics lies in its treatment of feedback as a generative force rather than a corrective mechanism. In traditional systems thinking, feedback loops are often categorized as either positive (amplifying change) or negative (stabilizing change), with an emphasis on using negative feedback to achieve equilibrium. However, Neodynamics recognizes that feedback plays a far more nuanced role in fostering adaptability. Positive feedback, for instance, can drive systems toward new equilibria or even enable the emergence of entirely novel structures by amplifying exploratory behaviors. This approach mirrors the concept of autopoiesis in biological systems, where self-reinforcing processes enable organisms to create and maintain their own internal coherence (Maturana & Varela, 1980). Neodynamics applies this principle to human-designed systems, treating feedback not as a tool for maintaining stability but as a dynamic driver of exploration and adaptation.

In practical terms, Neodynamics offers a structured means of designing feedback systems that promote continuous learning. For example, in the context of artificial intelligence, Neodynamic systems can incorporate recursive feedback loops that evaluate and refine model performance in real-time. Unlike static machine learning models that are constrained by their initial training data, such systems dynamically adjust their parameters and objectives based on evolving user input and environmental conditions. This enables the creation of AI systems that are not only more robust but also more relevant in diverse and shifting contexts, addressing a critical limitation of contemporary AI architectures (LeCun et al., 2015).

Similarly, Neodynamics introduces a paradigm shift in the design of social and organizational systems. Current frameworks often treat decision-making as a static process, relying on hierarchical structures and linear planning models. In contrast, Neodynamics embeds adaptability into decision-making processes through recursive choice mechanisms, allowing organizations to respond dynamically to both internal feedback and external changes. This is particularly relevant in industries such as logistics and supply chain management, where the ability to recalibrate operations in response to real-time disruptions can be the difference between system resilience and collapse (Ivanov & Dolgui, 2020). By emphasizing adaptability and emergent coherence, Neodynamics enables organizations to not only survive disruptions but also leverage them as opportunities for innovation and growth.

Neodynamics situates itself at the confluence of theoretical innovation and practical necessity. Its rejection of static paradigms, emphasis on transformative adaptation, and dynamic use of feedback represent a significant evolution in systems thinking. By integrating interdisciplinary insights into a cohesive framework, Neodynamics provides the tools needed to engineer systems capable of thriving in complexity, offering a vision for the future that aligns with the demands of a rapidly changing world.

The most significant contribution of Neodynamics to systems thinking is its emphasis on the Unified Field of Adaptive Potential (UFAP) and the SPARC Framework as operational tools. Unlike earlier paradigms, which often remain theoretical or descriptive, Neodynamics bridges the gap between theory and application by offering mechanisms to design, evaluate, and optimize adaptive systems.

The UFAP concept is particularly innovative in its ability to map the space of possible actions a system can take in response to changing conditions. This theoretical construct draws from the multidimensional probability spaces in quantum mechanics and information theory, where uncertainty and possibility coexist as defining properties (Schrödinger, 1944; Shannon, 1948). In Neodynamics, UFAP enables system designers to model not just the probable but also the possible, incorporating pathways that may appear improbable under current conditions but could become essential in an evolving environment. This ensures that systems maintain flexibility and are not constrained by past data or fixed assumptions, a limitation observed in traditional predictive models.

The SPARC Framework operationalizes UFAP by introducing a dual focus: the Spectrum of Possibility and Recursive Choice. The Spectrum of Possibility captures the range of outcomes a system can generate, ensuring diversity and adaptability in its responses. This diversity mirrors the biological principle of variation as a prerequisite for evolution, where a wide range of traits increases the likelihood of survival in changing environments (Darwin, 1859). Recursive Choice, on the other hand, introduces a feedback-driven process that continuously refines system decisions based on real-time feedback and emergent conditions. Together, these mechanisms transform static decision-making into an iterative, self-correcting process that aligns with the principles of complexity science (Holland, 1992).

Neodynamics also advances the discourse on emergence, a concept central to complex adaptive systems. In earlier frameworks, emergence is often treated as a descriptive phenomenon—patterns arising from the interactions of system components. Neodynamics goes further by emphasizing emergent coherence as an actionable principle. Emergent coherence occurs when a system's components self-organize into a functional whole, achieving alignment without external control. This principle is critical for designing systems that must function autonomously in unpredictable environments, such as decentralized AI networks or distributed governance systems.

For instance, in distributed AI, Neodynamics provides a framework for designing systems where individual agents can interact, adapt, and converge on coherent behaviors without relying on centralized control mechanisms. This is particularly relevant in applications like swarm robotics, where multiple autonomous units must coordinate in real-time to achieve collective objectives under dynamic conditions (Brambilla et al., 2013). Similarly, in governance, emergent coherence enables policy frameworks to adapt organically to shifting societal needs, avoiding the inefficiencies of rigid, top-down control structures.

By positioning itself as a unifying paradigm, Neodynamics draws strength from its ability to synthesize insights from diverse disciplines while addressing their limitations. It offers a path forward for the design of systems that are not only resilient but capable of continuous learning, adaptation, and evolution. This capacity to thrive in uncertainty is what makes Neodynamics uniquely suited to the challenges of the 21st century, where complexity and unpredictability are the defining characteristics of human and technological systems.

Historical and Conceptual Foundations
Origins of Systems Thinking

Systems thinking emerged in the mid-20th century as an effort to address the limitations of reductionist approaches to understanding complex phenomena. At its core, systems thinking posits that the behavior of a system cannot be fully understood by analyzing its individual components in isolation. Instead, it focuses on the relationships, interactions, and feedback loops that define a system's structure and dynamics (Bertalanffy, 1968). This shift in perspective laid the foundation for a broad range of disciplines, from ecology and cybernetics to organizational management and engineering.

Ludwig von Bertalanffy's General Systems Theory (GST) was a pivotal moment in the development of systems thinking. GST aimed to establish universal principles governing the behavior of systems, regardless of their specific domain. Bertalanffy introduced concepts such as open systems, which exchange energy and information with their environment, and emphasized the importance of feedback in maintaining system stability. These principles provided a framework for understanding how systems adapt and interact with their surroundings, offering early insights into the role of adaptability in complex environments.

However, GST was primarily descriptive and focused on equilibrium, assuming that systems naturally gravitate toward stable states. This assumption limited its applicability to dynamic and rapidly changing environments, where systems often operate far from equilibrium. Neodynamics builds on Bertalanffy's foundational work by shifting the focus from stability to adaptability. It argues that systems must not only respond to external perturbations but actively explore and reorganize to thrive under conditions of uncertainty.

The emergence of cybernetics in the 1940s further advanced systems thinking by introducing the concept of feedback loops as mechanisms for control and communication within systems. Norbert Wiener's work in cybernetics emphasized self-regulation, where systems use feedback to adjust their behavior and achieve desired outcomes (Wiener, 1948). This principle has had profound implications for fields ranging from automated control systems to biological models of homeostasis. However, traditional cybernetics often prioritized maintaining pre-determined objectives over fostering emergent adaptability. Neodynamics extends cybernetic principles by integrating feedback not just as a tool for stabilization but as a driver for exploration and transformation.

The development of systems ecology in the 1960s brought a new dimension to systems thinking by emphasizing the interconnectedness of ecological networks. Researchers such as Howard T. Odum explored energy flows, nutrient cycling, and feedback loops within ecosystems, illustrating how these processes sustain life (Odum, 1969). Systems ecology demonstrated that ecosystems are inherently adaptive, capable of reorganizing in response to disturbances. This ecological perspective deeply influences Neodynamics, particularly its emphasis on emergent coherence and the role of diversity in fostering resilience.

Another critical development in systems thinking was the recognition of complexity as a defining characteristic of many systems. Complex systems are characterized by nonlinear interactions, emergent properties, and unpredictability, challenging traditional reductionist approaches. The advent of complexity science in the 1980s and 1990s, spearheaded by researchers like John H. Holland and Murray Gell-Mann, provided new tools for understanding these systems (Holland, 1992; Gell-Mann, 1994). Complexity science introduced concepts such as adaptive agents, phase transitions, and self-organization, which are central to Neodynamics. While complexity science has been instrumental in describing the behavior of adaptive systems, Neodynamics goes further by offering actionable constructs—such as UFAP and SPARC—for engineering systems that can navigate complexity effectively.

The evolution of systems thinking demonstrates a progressive refinement of our understanding of complexity, adaptability, and feedback. From Bertalanffy's focus on equilibrium to the dynamic adaptability emphasized by complexity science, each development has contributed to the theoretical foundations of Neodynamics. By synthesizing these insights into a cohesive framework, Neodynamics positions itself as the next step in the evolution of systems thinking, offering both theoretical depth and practical tools for designing systems capable of thriving in a rapidly changing world.

Influences from Theoretical Disciplines

Neodynamics is deeply informed by a range of theoretical disciplines, each of which contributes unique insights into the nature of complexity, adaptability, and emergent behavior. By synthesizing these diverse perspectives, Neodynamics establishes a cohesive framework that addresses the limitations of traditional systems approaches while advancing the frontiers of adaptive design.

Thermodynamics: Beyond Equilibrium

The principles of thermodynamics, particularly those related to energy flows and entropy, are foundational to understanding how systems adapt and evolve. Traditional thermodynamics, as formulated in the 19th century, focused on closed systems that eventually reach equilibrium states. This perspective, while useful for understanding static systems, is insufficient for describing the behavior of open systems that exchange energy and information with their environment.

Ilya Prigogine's work on dissipative structures revolutionized this field by demonstrating that systems far from equilibrium can self-organize through the dissipation of energy (Prigogine, 1980). These dissipative structures exhibit emergent behaviors that enable them to adapt and maintain coherence under dynamic conditions.

This principle directly informs Neodynamics' emphasis on emergent coherence, where systems reorganize themselves through iterative processes of energy and information exchange. By extending thermodynamic principles to the realm of complex, adaptive systems, Neodynamics positions itself as a framework for understanding and engineering systems that thrive far from equilibrium.

Information Theory: Managing Uncertainty

Claude Shannon's information theory provides a mathematical foundation for understanding how systems process, transmit, and manage information in the presence of uncertainty (Shannon, 1948). Central to this theory is the concept of entropy, which quantifies the uncertainty or disorder within a system. In traditional applications, information theory has been used to optimize communication channels, ensuring that signals are transmitted efficiently and accurately.

Neodynamics draws on these principles but reorients their focus from static optimization to dynamic adaptability. The Unified Field of Adaptive Potential (UFAP) extends Shannon's entropy by conceptualizing a multidimensional space of possibilities that systems can explore in response to changing conditions. This construct enables systems to not only reduce uncertainty but to leverage it as a driver for exploration and innovation. By integrating information theory with principles of feedback and emergence, Neodynamics provides a framework for managing complexity in dynamic environments.

Quantum Mechanics: Probabilistic States and Adaptation

The probabilistic nature of quantum mechanics offers valuable analogies for understanding adaptability in complex systems. In quantum mechanics, particles exist in superpositions, occupying multiple states simultaneously until observed or measured (Heisenberg, 1927). This principle challenges deterministic notions of system behavior, emphasizing the role of probability and uncertainty in shaping outcomes.

Neodynamics incorporates similar principles through UFAP, which conceptualizes the adaptive potential of a system as a probabilistic landscape. Rather than committing to a single course of action, systems governed by Neodynamic principles explore multiple pathways simultaneously, dynamically recalibrating based on feedback and emergent conditions. This probabilistic approach enables systems to remain flexible and responsive, avoiding the rigidity that characterizes traditional optimization frameworks.

Game Theory: Strategic Adaptation

Game theory, which analyzes strategic interactions among agents, provides critical insights into how systems navigate competitive and cooperative dynamics. Traditional game theory assumes rational agents acting within defined payoff structures, seeking to optimize their outcomes under fixed rules. While this approach has been widely applied in economics, political science, and biology, its reliance on static equilibria limits its applicability to dynamic and evolving environments.

Neodynamics incorporates principles from evolutionary game theory, which extends the traditional framework by allowing strategies to evolve over time in response to interactions and feedback. In this context, the SPARC Framework's Recursive Choice mechanism parallels the iterative refinement of strategies observed in evolutionary dynamics. Agents within a Neodynamic system continuously update their decisions based on feedback, fostering adaptability and emergent coherence. This capacity for real-time recalibration enables systems to thrive in competitive environments where conditions and objectives are constantly shifting.

Biology: Adaptation and Emergence

Biological systems, from individual organisms to ecosystems, exemplify the principles of adaptability and emergent behavior that are central to Neodynamics. Evolutionary processes, driven by variation, selection, and heredity, provide a natural model for understanding how systems develop resilience and complexity over time (Darwin, 1859). These processes highlight the importance of diversity as a prerequisite for adaptability, ensuring that systems can explore a wide range of responses to environmental changes.

Neodynamics builds on these insights by embedding diversity within the Spectrum of Possibility, a core component of the SPARC Framework. This diversity allows systems to generate novel solutions and reorganize dynamically in response to feedback. Additionally, biological systems demonstrate the power of distributed processes in achieving emergent coherence. For example, ant colonies and neural networks self-organize through local interactions among individual agents, producing collective behaviors that exceed the capabilities of any single component. Neodynamics adopts similar principles, emphasizing the role of distributed decision-making in fostering adaptability and coherence across scales.

Ecology: Resilience and Self-Organization

The field of ecology offers further insights into the adaptive capacity of systems operating within interconnected networks. Ecosystems are inherently adaptive, characterized by feedback loops that regulate energy flows, resource availability, and population dynamics (Odum, 1969). These feedback mechanisms enable ecosystems to absorb disturbances, reorganize, and achieve new states of equilibrium—an attribute known as ecological resilience.

Neodynamics extends this concept by emphasizing transformative adaptation, where systems not only recover from disturbances but evolve to meet new challenges. This perspective aligns with the ecological principle of succession, where ecosystems reorganize into more complex configurations following a disruption. By integrating these ecological principles, Neodynamics provides a framework for designing human systems that emulate the adaptability and resilience of natural ecosystems.

Sociology and Organizational Theory: Dynamic Structures

Sociological and organizational theories contribute to Neodynamics by exploring how social systems adapt to change and maintain coherence in dynamic environments. Research on organizational behavior highlights the importance of feedback, distributed leadership, and flexible structures in enabling organizations to respond effectively to external pressures (Mintzberg, 1979). Traditional hierarchical models, while efficient under stable conditions, often fail in the face of uncertainty and complexity.

Neodynamics advocates for decentralized, adaptive structures that mirror the distributed processes observed in natural and artificial systems. The SPARC Framework's Recursive Choice mechanism facilitates this by embedding feedback loops at all levels of decision-making, ensuring that organizations remain agile and responsive. This approach fosters not only resilience but also innovation, enabling systems to leverage complexity as an opportunity rather than a threat.

Economics: Complexity and Adaptive Markets

Economics has long grappled with the challenge of understanding and predicting the behavior of complex market systems. Traditional economic models, rooted in classical and neoclassical theories, often assume equilibrium, rational agents, and linear causality. These assumptions have been widely criticized for their inability to account for the dynamic, nonlinear interactions that characterize real-world markets (Farmer & Foley, 2009). Financial crises, such as the 2008 collapse, highlighted the failure of static models to anticipate systemic risks created by interdependencies and feedback loops.

Neodynamics offers a framework for addressing these limitations by incorporating concepts of complexity and adaptability into economic systems. The Unified Field of Adaptive Potential (UFAP) can be applied to model the space of potential market responses to shocks, enabling policymakers and analysts to explore scenarios beyond the constraints of equilibrium theory. The SPARC Framework further extends this adaptability by embedding recursive decision-making processes within economic models, allowing systems to self-correct dynamically in response to emergent conditions. For example, adaptive monetary policies informed by Neodynamics could recalibrate in real time based on market feedback, minimizing systemic risks and fostering stability in volatile environments.

Engineering and Resilience Thinking

Engineering has traditionally focused on optimizing systems for efficiency and reliability under specific operating conditions. While effective in controlled environments, this approach often produces brittle systems that fail under unexpected stress. The concept of resilience engineering emerged as a response, emphasizing the ability of systems to absorb shocks and recover functionality (Hollnagel, Woods, & Leveson, 2006). Resilience engineering prioritizes robustness and flexibility, offering a foundation for designing systems that can handle uncertainty.

Neodynamics builds on resilience engineering by shifting the focus from recovery to continuous evolution. In this framework, systems are not merely designed to withstand shocks but to adapt and reorganize dynamically in response to change. The SPARC Framework's Spectrum of Possibility ensures that systems maintain a diverse range of responses, while Recursive Choice enables iterative refinement based on feedback. This approach is particularly relevant in fields such as infrastructure design, where adaptive systems can optimize resource allocation, enhance sustainability, and respond dynamically to environmental changes.

Physics: Nonlinearity and Chaos

Physics, particularly the study of nonlinear systems and chaos theory, provides additional insights into the behavior of complex systems. Chaos theory, as popularized by Lorenz's work on weather systems, reveals how small changes in initial conditions can produce vastly different outcomes, a phenomenon often referred to as the "butterfly effect" (Lorenz, 1963). This sensitivity to initial conditions highlights the inherent unpredictability of dynamic systems, challenging traditional deterministic models.

Neodynamics incorporates these principles by acknowledging the limitations of predictability in complex environments. Rather than attempting to control or predict outcomes, systems designed within the Neodynamic framework leverage feedback and exploration to navigate uncertainty. The UFAP concept aligns closely with phase space in physics, representing the multidimensional landscape of possible states a system can occupy. By embedding adaptability into system design, Neodynamics transforms unpredictability from a constraint into a driver of innovation and resilience.

Interdisciplinary Synthesis

The influences from these theoretical disciplines converge in Neodynamics, creating a comprehensive framework that transcends traditional boundaries. By integrating insights from thermodynamics, information theory, biology, sociology, economics, and physics, Neodynamics provides both the theoretical foundation and practical tools needed to design systems capable of thriving in complexity. This interdisciplinary synthesis positions Neodynamics not as a replacement for existing paradigms but as an evolution that addresses their limitations while building on their strengths. Through constructs like UFAP and SPARC, it transforms abstract principles into actionable mechanisms, enabling the design of systems that are not only resilient but dynamically adaptive.

Despite the significant advances made by systems theory, cybernetics, complexity science, and related disciplines, critical gaps remain in their ability to guide the design and operation of adaptive systems. These gaps include a lack of actionable tools for dynamic adaptability, an overemphasis on stability and equilibrium, and insufficient integration of feedback as a generative force. Neodynamics addresses these shortcomings by offering a unifying framework that not only describes but also enables the engineering of systems capable of thriving in uncertain and evolving environments.

Most existing frameworks, such as General Systems Theory and complex adaptive systems theory, provide descriptive models of how systems behave under various conditions. While these models are valuable for understanding interactions, feedback loops, and emergent behaviors, they often lack prescriptive tools for designing systems that embody these principles. For example, complexity science identifies the importance of agent-based interactions in producing emergent coherence but does not specify how to engineer such interactions in practical applications (Holland, 1992).

Neodynamics bridges this gap through its core constructs: the Unified Field of Adaptive Potential (UFAP) and the SPARC Framework. These constructs translate abstract principles into actionable methodologies, enabling system designers to map adaptive possibilities, implement feedback-driven decision-making, and foster emergent coherence. By shifting the focus from theoretical description to practical application, Neodynamics ensures that adaptability is not merely observed but embedded in the system's architecture.

Traditional systems approaches, such as those rooted in thermodynamics and cybernetics, often emphasize equilibrium and homeostasis as desirable states. While these principles are effective in stable environments, they are ill-suited to systems operating under conditions of rapid change and uncertainty. For example, cybernetic models of control typically aim to maintain stability through negative feedback, which suppresses deviations from a predefined norm (Wiener, 1948). While this approach can be effective in maintaining functional consistency, it limits the system's ability to explore novel pathways or reorganize in response to emerging challenges.

Neodynamics rejects the primacy of equilibrium, instead prioritizing adaptability and dynamic stability. Dynamic stability refers to a system's ability to maintain coherence while continuously recalibrating in response to external perturbations. This principle aligns with insights from Prigogine's work on dissipative structures, which demonstrates that systems can self-organize into new configurations when operating far from equilibrium (Prigogine, 1980). By incorporating this understanding, Neodynamics enables the design of systems that do not resist change but actively embrace it as a driver of evolution.

Feedback is a central concept in many systems frameworks, but its role is often limited to maintaining control or stability. Negative feedback, for instance, is typically used to suppress deviations, while positive feedback is often viewed as a destabilizing force to be minimized. This limited perspective fails to capture the transformative potential of feedback in driving adaptation and innovation.

Neodynamics reframes feedback as a generative force that enables systems to explore, learn, and evolve. Positive feedback, rather than being inherently destabilizing, is recognized as a mechanism for amplifying exploratory behaviors and fostering emergent coherence. Similarly, recursive feedback processes, as operationalized in the SPARC Framework, enable systems to iteratively refine their decisions, integrating new information and recalibrating objectives in real time. This approach ensures that feedback is not merely a corrective tool but a foundational mechanism for adaptability and growth.

Existing systems theories often focus on specific scales or domains, limiting their applicability to complex, interconnected environments. For example, resilience engineering primarily addresses infrastructure and disaster response, while complexity science is often confined to biological or economic systems. This compartmentalization creates gaps in understanding how adaptive principles can be applied across diverse contexts.

Neodynamics addresses this limitation by providing a unifying framework that integrates adaptability across scales and domains. Constructs like UFAP and SPARC are designed to be universally applicable, whether in the context of artificial intelligence, governance, healthcare, or ecological management. By emphasizing shared principles of adaptability, feedback, and emergent coherence, Neodynamics offers a cohesive approach to designing systems that operate effectively in both local and global contexts.

Perhaps the most significant gap addressed by Neodynamics is the absence of frameworks that prioritize transformative adaptation. Traditional approaches often focus on resilience, defined as the ability to absorb shocks and return to a previous state. While resilience is valuable, it assumes that the prior state remains desirable or viable, an assumption that does not hold in rapidly changing environments.

Neodynamics moves beyond resilience by advocating for systems that not only recover from disruptions but evolve to meet new challenges. This perspective is particularly relevant in addressing global crises, such as climate change and technological disruption, where static solutions are insufficient. By embedding transformative adaptation into its constructs, Neodynamics provides a pathway for systems to continuously expand their adaptive potential, ensuring their relevance and functionality in an uncertain future.

In addressing these gaps, Neodynamics establishes itself as a necessary evolution in systems thinking. It synthesizes the strengths of existing frameworks while overcoming their limitations, offering a comprehensive and actionable approach to designing systems that thrive in complexity.

Core Principles of Neodynamics
Adaptability and Dynamic Stability

Adaptability and dynamic stability are the foundational principles of Neodynamics, reflecting its focus on systems that thrive under conditions of uncertainty and complexity. Traditional systems design often prioritizes stability, defining success as the ability to maintain a predefined state or function. While this approach is effective in controlled environments, it fails to account for the dynamic and often unpredictable nature of real-world systems. Neodynamics reframes stability not as a static condition but as a dynamic process of continuous adaptation, enabling systems to evolve in response to change.

Adaptability refers to a system's capacity to respond effectively to internal or external changes without losing its functional coherence. This involves more than simply reacting to perturbations; true adaptability requires systems to anticipate potential disruptions, explore alternative pathways, and reorganize themselves dynamically to align with shifting objectives. In biological terms, adaptability is akin to the evolutionary process, where variation, selection, and heredity enable organisms to survive and thrive in changing environments (Darwin, 1859).

Neodynamics operationalizes adaptability through the Unified Field of Adaptive Potential (UFAP), which maps the range of actions a system can take to maintain or enhance its functionality. By embedding adaptability into system design, Neodynamics ensures that systems remain flexible, resilient, and capable of exploring novel solutions in the face of uncertainty. This approach shifts the focus from optimizing for specific conditions to optimizing for the ability to adapt across a wide range of scenarios.

Traditional systems often rely on static stability, where feedback mechanisms are used to suppress deviations from a predefined equilibrium. This approach assumes that stability is synonymous with success, leading to systems that resist change rather than embrace it. For example, infrastructure systems designed for static stability—such as dams optimized for historical weather patterns—are increasingly vulnerable to the variability introduced by climate change. When these systems encounter conditions beyond their design parameters, they often fail catastrophically.

Dynamic stability, by contrast, recognizes that stability is not a fixed state but a process of continuous recalibration. Systems exhibiting dynamic stability maintain coherence by adapting to change rather than resisting it. This principle is evident in natural ecosystems, where feedback loops and distributed processes enable organisms to self-organize and achieve new equilibria after disruptions. For instance, coral reefs damaged by bleaching events often reorganize into alternative states that sustain biodiversity, albeit in a different form (Hughes et al., 2003).

In Neodynamics, dynamic stability is achieved through recursive feedback processes that allow systems to monitor, evaluate, and adjust their behavior in real time. The SPARC Framework's Recursive Choice mechanism operationalizes this principle, enabling systems to refine their decisions iteratively based on feedback and emergent conditions. This ensures that systems remain coherent and functional even as their environments evolve.

The principles of adaptability and dynamic stability have broad applications across diverse domains. In artificial intelligence, for instance, static models trained on fixed datasets often struggle to generalize when exposed to new or unanticipated conditions.

Neodynamics addresses this limitation by embedding feedback-driven adaptability into AI architectures, enabling systems to recalibrate dynamically based on real-time input. This approach not only improves performance but also enhances the relevance and coherence of AI-generated outputs in complex environments.

In governance, adaptability and dynamic stability inform the design of policies and institutions capable of evolving in response to societal and environmental changes. Traditional governance models often rely on rigid hierarchies and long-term plans that fail to account for the uncertainty inherent in crises such as pandemics or climate disasters. By integrating Neodynamic principles, governance systems can incorporate feedback loops and real-time data analysis, enabling policies to be continuously evaluated and adjusted to meet shifting needs. By prioritizing these principles, Neodynamics provides a framework for designing systems that not only survive disruptions but leverage them as opportunities for growth and innovation.

Biological systems provide some of the most compelling examples of dynamic stability in action. Organisms and ecosystems alike operate in environments that are inherently variable, requiring continuous adjustment to survive. One illustrative example is the process of homeostasis, where organisms maintain internal stability (such as body temperature or pH levels) despite fluctuations in external conditions. This stability is not achieved through rigidity but through dynamic feedback mechanisms that allow for real-time recalibration (Cannon, 1932). For instance, thermoregulation in mammals involves both immediate responses, such as shivering or sweating, and longer-term adaptations, such as acclimatization to seasonal changes.

At the ecosystem level, dynamic stability is evident in predator-prey relationships, where population sizes oscillate in response to resource availability and interspecies interactions. These oscillations are not indicative of instability; rather, they reflect the system's ability to self-correct and maintain functional coherence over time (Lotka, 1925; Volterra, 1926). Neodynamics draws directly from these natural processes, embedding similar feedback-driven recalibration mechanisms into its constructs to ensure that artificial and organizational systems exhibit comparable adaptability.

Dynamic stability is equally relevant to human-designed systems, where it provides a framework for addressing challenges in infrastructure, governance, and technology. For instance, urban water management systems designed around static assumptions of rainfall patterns are increasingly overwhelmed by the variability introduced by climate change. Flooding events and droughts expose the fragility of these systems, which are optimized for historical averages rather than future uncertainties. A Neodynamic approach would embed adaptability into these systems through real-time monitoring, predictive analytics, and dynamic resource allocation. By leveraging feedback from sensors and environmental models, water systems could recalibrate their operations continuously, maintaining stability even under extreme conditions.

In governance, dynamic stability enables institutions to evolve in response to societal and environmental changes. Traditional governance models often rely on rigid hierarchies and long-term plans that are difficult to adjust once implemented. This rigidity creates vulnerabilities when conditions shift, as seen during crises such as the COVID-19 pandemic. Neodynamics offers an alternative by emphasizing adaptive governance, where policies are continuously evaluated and revised based on feedback from stakeholders and real-time data. This iterative process ensures that governance systems remain aligned with public needs and emerging challenges, fostering resilience and trust.

The Unified Field of Adaptive Potential (UFAP) is central to Neodynamics' approach to dynamic stability. By conceptualizing the space of possible actions a system can take, UFAP enables designers to identify and prioritize pathways that maximize adaptability. This probabilistic landscape allows systems to explore multiple responses simultaneously, avoiding the rigidity of traditional optimization frameworks.

The SPARC Framework operationalizes these principles through its dual components: the Spectrum of Possibility and Recursive Choice. The Spectrum of Possibility ensures that systems maintain a diverse range of responses, mirroring the role of genetic variation in biological evolution. Recursive Choice introduces iterative feedback loops that refine system decisions in real time, enabling continuous recalibration. Together, these constructs transform adaptability from a reactive process into a proactive strategy for maintaining stability under dynamic conditions.

The integration of adaptability and dynamic stability into system design marks a departure from traditional engineering and planning paradigms. Neodynamic systems prioritize flexibility and resilience over narrow optimization, ensuring that they remain functional and coherent even in the face of uncertainty. This approach is particularly relevant in the context of global challenges such as climate change, technological disruption, and geopolitical instability. By embedding principles of dynamic stability into artificial intelligence, infrastructure, and governance, Neodynamics provides a pathway for creating systems that not only survive disruptions but thrive within them.

Emergent Coherence

Emergent coherence is a defining principle of Neodynamics, capturing the phenomenon where systems composed of diverse, interacting components self-organize into functional wholes. Unlike coherence imposed by external control or rigid hierarchies, emergent coherence arises naturally from the interactions within the system. This principle is central to Neodynamics, as it enables systems to maintain alignment and purpose even in highly dynamic and unpredictable environments.

Emergent coherence differs fundamentally from top-down coordination. In traditional systems, coherence is often achieved through centralized control mechanisms, which dictate behavior and enforce alignment among components. While effective in stable conditions, such approaches fail in dynamic environments where centralized models cannot process information or implement adjustments rapidly enough to maintain functionality. By contrast, emergent coherence leverages decentralized interactions and local decision-making, enabling systems to adapt and reorganize without requiring a central authority.

In natural systems, emergent coherence is ubiquitous. Ant colonies, for instance, exhibit remarkable coordination despite the absence of centralized control. Individual ants follow simple behavioral rules, such as depositing pheromone trails, which collectively produce complex behaviors such as foraging or nest building (Gordon, 1999). Similarly, neural networks in the brain achieve coherence through the interactions of billions of neurons, each processing local inputs and outputs. This self-organizing capability is essential for dynamic functions such as learning and memory.

Neodynamics draws inspiration from these natural systems, embedding principles of emergent coherence into its constructs. The Unified Field of Adaptive Potential (UFAP) provides a framework for mapping the potential interactions among components, while the SPARC Framework ensures that these interactions are iterative and feedback-driven. Together, these mechanisms allow systems to achieve alignment and coherence even in the absence of centralized oversight.

Emergent coherence arises from three primary mechanisms: diversity, feedback, and local interaction. Diversity ensures that systems have a broad range of potential responses to explore, increasing the likelihood of discovering adaptive solutions. Feedback channels the system's behavior toward coherence by reinforcing successful interactions and attenuating less effective ones. Local interactions enable components to respond to immediate conditions, fostering flexibility and scalability.

The SPARC Framework operationalizes these mechanisms by incorporating the Spectrum of Possibility and Recursive Choice into system design. The Spectrum of Possibility ensures that systems maintain a diversity of responses, preventing premature convergence on suboptimal solutions. Recursive Choice facilitates the iterative refinement of interactions, allowing systems to dynamically adjust their behavior based on emerging patterns of coherence. These processes mirror the principles of natural selection, where variation, selection, and heredity drive the evolution of adaptive traits.

Emergent coherence has broad applications across domains. In artificial intelligence, it informs the design of distributed systems such as swarm robotics and decentralized machine learning models. For example, in swarm robotics, individual robots follow simple rules for movement and communication, enabling the swarm to achieve collective objectives such as search-and-rescue operations or environmental monitoring. This decentralized approach ensures that the system remains flexible and robust, even when individual components fail or conditions change.

In governance, emergent coherence enables the development of policies and institutions that adapt organically to societal needs. Traditional governance models often rely on top-down coordination, which is slow to respond to crises or changing public demands. By embedding feedback loops and decentralized decision-making into governance structures, Neodynamics fosters adaptive alignment among stakeholders, enabling policies to evolve dynamically in response to real-time data and public input.

In healthcare, emergent coherence can be applied to systems for personalized medicine and public health. For instance, decentralized networks of wearable health monitors can self-organize to detect population-level trends, enabling early interventions for epidemics. Similarly, feedback-driven algorithms can ensure that treatments are continuously adjusted to align with individual patient responses, enhancing effectiveness and reducing side effects.

While emergent coherence offers significant advantages, it also presents challenges in system design. Achieving coherence through decentralized interactions requires careful calibration of feedback mechanisms to prevent instability or overconvergence on suboptimal patterns. Additionally, designing systems that balance diversity and coherence remains a complex task, as excessive diversity can lead to fragmentation while insufficient diversity reduces adaptability.

Neodynamics addresses these challenges by providing a structured framework for embedding the principles of emergent coherence into system design. By leveraging UFAP and SPARC, it ensures that systems maintain a balance between flexibility and alignment, enabling them to thrive in dynamic and uncertain environments. This approach transforms emergent coherence from a descriptive concept into an actionable principle, equipping designers with the tools needed to create adaptive and resilient systems.

Feedback-Driven Processes

Feedback-driven processes are the backbone of Neodynamics, enabling systems to continuously adjust and refine their behavior in response to internal and external changes. Unlike static systems, which operate on predetermined rules or fixed optimization goals, feedback-driven systems thrive by integrating information from their environment and using it to recalibrate dynamically. This iterative process of feedback integration is essential for maintaining adaptability, coherence, and stability in complex, interconnected environments.

Feedback is a fundamental mechanism through which systems monitor their performance, detect deviations, and implement corrections. It can be broadly categorized into two types: negative feedback, which stabilizes systems by counteracting deviations from a desired state, and positive feedback, which amplifies specific behaviors or trends to drive transformation. While traditional systems approaches often emphasize negative feedback for maintaining control and stability, Neodynamics expands this scope by treating feedback as a generative force that fosters learning, exploration, and innovation.

For example, in biological systems, feedback mechanisms regulate processes ranging from cellular metabolism to ecosystem dynamics. Homeostasis in organisms relies on negative feedback loops to maintain internal stability, such as regulating blood sugar levels through insulin release. Conversely, positive feedback drives transformative processes, such as the amplification of gene expression during cellular differentiation or the cascading dynamics of predator-prey relationships that shape ecosystems over time (Lotka, 1925; Odum, 1969).

Neodynamics incorporates both types of feedback into its constructs, ensuring that systems can balance stability with adaptability. The SPARC Framework's Recursive Choice mechanism is particularly suited for integrating feedback, as it allows systems to iteratively evaluate their actions, refine their strategies, and adjust their objectives based on emergent conditions.

Recursive feedback is a defining feature of Neodynamics, enabling systems to continuously refine their behavior through iterative cycles of evaluation and adjustment. This process mirrors the principles of machine learning, where models improve their performance by analyzing errors and updating their parameters. However, Neodynamics extends this concept beyond algorithmic optimization, embedding recursive feedback into the structure of entire systems to ensure adaptability at all levels.

In artificial intelligence, recursive feedback facilitates the design of systems that adapt dynamically to changing inputs. For instance, conversational AI models equipped with feedback loops can refine their responses in real time, improving their relevance and coherence based on user interactions. Similarly, in autonomous vehicles, recursive feedback enables continuous learning from sensor data, allowing the system to adjust its navigation strategies in response to new environmental conditions.

In organizational systems, recursive feedback ensures that decision-making processes remain agile and responsive. Traditional hierarchical models often suffer from delays and inefficiencies, as information must pass through multiple levels before decisions are implemented. By embedding feedback loops at all levels of the organization, Neodynamics enables real-time adjustments that align with evolving goals and conditions. This approach is particularly valuable in crisis management, where timely responses are critical for mitigating risks and maintaining operational continuity.

Feedback-driven processes are integral to fostering emergent coherence in systems composed of diverse and interacting components. Positive feedback amplifies successful interactions, reinforcing patterns that align with system-wide objectives. Negative feedback, meanwhile, prevents overconvergence on suboptimal patterns by introducing corrective forces. Together, these mechanisms create a dynamic balance that allows systems to self-organize and maintain coherence even in rapidly changing environments.

For example, in swarm robotics, individual robots rely on local feedback to coordinate their actions, enabling the swarm to achieve collective objectives without centralized control. By iteratively refining their behaviors based on interactions with other robots and environmental cues, the swarm achieves emergent coherence that is robust to disruptions and adaptable to new tasks.

The integration of feedback-driven processes has transformative potential across domains. In healthcare, feedback loops can enable personalized medicine by continuously monitoring patient responses to treatments and adjusting protocols in real time. Wearable devices that track physiological data provide the feedback necessary for dynamic treatment plans, ensuring that interventions remain effective as conditions evolve.

In governance, feedback-driven processes support the development of adaptive policies that respond dynamically to societal needs. Real-time data from sensors, social media, and citizen engagement platforms can inform policy adjustments, enabling governments to remain agile and responsive in the face of crises or public demand shifts.

In supply chains, feedback mechanisms facilitate real-time adjustments to production, inventory, and logistics. For example, during the COVID-19 pandemic, feedback-driven systems enabled some organizations to rapidly reconfigure their supply chains to meet changing demands for essential goods. These systems demonstrated resilience and adaptability by integrating information from across the supply chain and recalibrating operations continuously.

Despite their advantages, feedback-driven processes pose challenges in system design. Excessive reliance on feedback can lead to instability, as systems become overly sensitive to noise or short-term fluctuations. Similarly, poorly calibrated feedback mechanisms can result in oscillatory or chaotic behaviors that undermine coherence and stability. Neodynamics addresses these challenges by embedding feedback mechanisms within structured frameworks like UFAP and SPARC, ensuring that feedback is both actionable and context-aware.

By treating feedback as a dynamic and generative force, Neodynamics transforms it from a corrective tool into a driver of adaptability and innovation. This approach enables systems to not only respond to change but to leverage it as an opportunity for continuous improvement and evolution. Feedback-driven processes thus form the foundation for designing systems that thrive in complexity, ensuring their relevance and resilience in an unpredictable world.

Key Constructs of Neodynamics
Unified Field of Adaptive Potential (UFAP)

The Unified Field of Adaptive Potential (UFAP) is a theoretical construct at the heart of Neodynamics, representing the multidimensional space of possible actions and responses available to a system in a given context. By framing adaptability as a function of dynamic exploration within this field, UFAP provides a foundation for modeling and designing systems capable of thriving under uncertainty and complexity. This construct integrates insights from physics, information theory, and evolutionary biology, offering both theoretical depth and practical utility.

UFAP conceptualizes adaptability as a probabilistic landscape where each point represents a potential state or action that a system can adopt. Unlike static models that define fixed pathways or outcomes, UFAP accommodates the fluid nature of real-world systems by treating adaptability as an ongoing process of exploration, evaluation, and optimization. This dynamic perspective aligns with principles from quantum mechanics, where systems exist in superpositions of states until measured or observed (Heisenberg, 1927). Similarly, within UFAP, systems navigate a spectrum of possibilities, continuously recalibrating based on feedback and emergent conditions.

The key properties of UFAP include:

1.   Dynamism: The field evolves over time as the system interacts with its environment, reflecting changes in constraints, objectives, and opportunities.

2.   Multidimensionality: UFAP   captures   multiple dimensions of adaptation, including physical, informational, and organizational responses.

3.   Probabilistic Structure: Each potential state within the field is associated with a probability that reflects its feasibility, effectiveness, or desirability.

Formally, UFAP can be represented as a multidimensional probability density function (PDF), where represents a vector of possible actions or states. The shape of evolves dynamically as the system integrates feedback and recalibrates its objectives. This evolution can be modeled using differential equations that account for factors such as environmental changes, resource availability, and system constraints.

For example, consider a system navigating a two-dimensional UFAP, where represents resource allocation strategies and represents policy adjustments. The probability density function evolves according to the feedback-driven flow of probabilities and is a diffusion term that captures exploratory behavior. This formulation ensures that the system continuously evaluates potential actions while adapting to changes in its environment.

UFAP provides a versatile tool for modeling adaptability across a range of domains:

• Artificial Intelligence: In reinforcement learning, UFAP can guide the exploration of action spaces, enabling agents to identify optimal strategies in complex and uncertain environments. For instance, an autonomous vehicle navigating a dynamic traffic scenario can use UFAP to evaluate potential maneuvers, balancing safety, efficiency, and adaptability.

• Healthcare: UFAP can model the space of treatment options available to a physician, integrating patient-specific data, medical knowledge, and real-time feedback to identify personalized therapeutic pathways.

• Governance: Policy-makers can use UFAP to explore adaptive responses to crises, such as natural disasters or economic disruptions. By modeling the space of potential interventions, UFAP enables the identification of strategies that maximize resilience and societal well-being.

A critical feature of UFAP is its role in fostering emergent coherence. By enabling systems to explore diverse pathways, UFAP ensures that they maintain alignment with their overarching objectives while adapting to local and global changes. This process mirrors the role of diversity and selection in biological evolution, where populations explore a wide range of traits to ensure survival under changing conditions (Darwin, 1859). In Neodynamics, UFAP operationalizes this principle by embedding adaptability directly into the design of artificial and organizational systems.

While UFAP offers significant advantages, its implementation poses challenges. Modeling a high-dimensional adaptive landscape requires sophisticated computational tools and data integration capabilities. Additionally, the probabilistic nature of UFAP necessitates robust mechanisms for managing uncertainty, as overly conservative or aggressive exploration can lead to inefficiencies or instability.

To address these challenges, Neodynamics integrates UFAP with the SPARC Framework, which provides structured processes for navigating the adaptive landscape. By combining the exploratory potential of UFAP with the iterative refinement enabled by Recursive Choice, Neodynamics ensures that systems leverage the full spectrum of possibilities while maintaining coherence and functionality.

The Unified Field of Adaptive Potential is a transformative construct that redefines how adaptability is conceptualized and implemented in complex systems. By framing adaptability as a dynamic process of exploration within a probabilistic landscape, UFAP provides a powerful tool for designing systems that thrive in uncertainty. Its integration with Neodynamics' broader framework ensures that adaptability becomes a foundational principle, enabling systems to evolve, innovate, and succeed in a rapidly changing world.

SPARC Framework (Spectrum of Possibility and Recursive Choice)

The SPARC Framework is a cornerstone of Neodynamics, operationalizing the principles of adaptability and emergent coherence by embedding feedback-driven processes into system design. SPARC stands for Spectrum of Possibility and Recursive Choice, emphasizing the dual mechanisms that enable systems to explore a wide range of potential responses and iteratively refine their decisions in real time. Together, these components ensure that systems remain adaptable, resilient, and coherent under dynamic conditions.

The Spectrum of Possibility represents the range of potential states, actions, or outcomes available to a system at any given moment. This diversity is essential for fostering adaptability, as it allows systems to explore multiple pathways in response to changing conditions. The concept draws from evolutionary biology, where genetic diversity provides the raw material for adaptation, and from information theory, where entropy represents the uncertainty or potential within a system (Shannon, 1948).

In the SPARC Framework, the Spectrum of Possibility is modeled as a multidimensional space, similar to the Unified Field of Adaptive Potential (UFAP). Each dimension corresponds to a variable or parameter that the system can adjust, such as resource allocation, policy choices, or behavioral strategies. The exploration of this space is guided by feedback loops, which evaluate the effectiveness of different pathways and reinforce those that align with the system's objectives.

Key properties of the Spectrum of Possibility include:

1. Breadth: The range of potential responses, ensuring that systems maintain flexibility and avoid overconvergence on suboptimal solutions.

2. Depth: The granularity of exploration, allowing systems to refine specific pathways while maintaining a broader perspective.

3. Dynamism: The ability of the spectrum to evolve as the system integrates feedback and recalibrates its objectives.

Recursive Choice is the second component of the SPARC Framework, providing a structured process for iteratively refining decisions based on feedback. Unlike traditional decision-making models, which often rely on static optimization or one-time evaluations, Recursive Choice treats decision-making as a continuous cycle of exploration, evaluation, and adjustment. This iterative approach enables systems to adapt dynamically, incorporating new information and responding to emergent conditions in real time.

Recursive Choice is governed by three key principles:

1.   Feedback Integration: Decisions are continuously evaluated based on the outcomes they produce, with feedback loops guiding subsequent iterations.

2.   Iterative Refinement: Systems do not commit to a single course of action but iteratively adjust their strategies to align with evolving objectives and constraints.

3.   Self-Correction: Errors or inefficiencies are treated as opportunities for learning and recalibration, ensuring that the system improves over time.

Mathematically, Recursive Choice can be modeled using optimization algorithms that incorporate feedback into their updates. For instance, gradient-based methods in machine learning adjust model parameters iteratively based on error signals, mirroring the principles of Recursive Choice. In Neodynamics, this process extends beyond algorithmic optimization to encompass the broader decision-making processes within systems.

The SPARC Framework has broad applicability across domains:

• Artificial Intelligence: SPARC enables the design of AI systems that adapt dynamically to changing inputs and objectives. For example, reinforcement learning agents guided by SPARC principles can explore diverse action spaces while iteratively refining their policies based on environmental feedback.

• Governance: In policymaking, the Spectrum of Possibility allows decision-makers to explore a range of potential interventions, while Recursive Choice ensures that policies are continuously evaluated and adjusted based on real-time data. This approach fosters adaptive governance, enabling institutions to respond effectively to crises and shifting societal needs.

• Healthcare: SPARC informs the development of personalized treatment protocols, where the Spectrum of Possibility represents the range of therapeutic options and Recursive Choice ensures that treatments are dynamically adjusted based on patient responses.

• Supply Chains: In logistics, SPARC facilitates the optimization of resource allocation and distribution strategies by integrating feedback from real-time demand and supply conditions.

The SPARC Framework plays a central role in fostering emergent coherence within systems. By maintaining a diverse spectrum of possibilities, SPARC ensures that systems do not prematurely converge on suboptimal solutions. Recursive Choice reinforces interactions that align with system-wide objectives, guiding the system toward coherence through iterative refinement. This balance of exploration and alignment mirrors the dynamics of natural systems, where local interactions and feedback loops drive the emergence of coherent behaviors.

Implementing the SPARC Framework requires careful calibration to balance exploration and exploitation. Excessive exploration can lead to inefficiencies, while overexploitation risks locking the system into suboptimal patterns. Neodynamics addresses these challenges by integrating SPARC with UFAP, ensuring that the exploration of possibilities is guided by probabilistic insights and feedback-driven processes.

The computational and organizational demands of SPARC also require robust data integration and analytical tools. For example, modeling the Spectrum of Possibility in high-dimensional spaces necessitates advanced machine learning and simulation techniques. Similarly, implementing Recursive Choice in governance or healthcare systems requires seamless feedback mechanisms that capture and analyze data in real time.

The SPARC Framework is a transformative tool for operationalizing adaptability and coherence in complex systems. By combining the Spectrum of Possibility with Recursive Choice, SPARC enables systems to navigate uncertainty, integrate feedback, and evolve dynamically. This approach ensures that systems remain functional, resilient, and aligned with their objectives, even in the face of rapid change. As a core construct of Neodynamics, SPARC exemplifies the principles of adaptability, feedback, and emergent coherence that define this new paradigm.

Applications of Neodynamics
Generative AI

The application of Neodynamics to generative artificial intelligence (AI) addresses critical challenges in ensuring adaptability, coherence, and scalability in systems that produce text, images, code, or other creative outputs. Generative AI systems often operate in dynamic and unpredictable environments where user needs, inputs, and contextual factors evolve continuously. Traditional approaches to designing such systems rely on static optimization of training data and model architectures, which limits their ability to adapt in real time. Neodynamics provides a framework for embedding adaptability and emergent coherence directly into the architecture of generative AI, enabling these systems to respond dynamically to changing conditions while maintaining functional stability.

Generative AI systems, such as GPT-based models or image generators like DALL-E, face several limitations rooted in static design principles:

1.   Coherence Across Outputs: Maintaining logical and contextual consistency across extended outputs remains a challenge,

particularly in text generation, where models must align local decisions (e.g., word choice) with broader objectives (e.g., thematic consistency).

2.   Adaptability to Novel Contexts: Models trained on static datasets struggle to generalize to novel scenarios or inputs that deviate significantly from their training distribution.

3.   Feedback Integration: Traditional generative systems lack robust mechanisms for incorporating user feedback or contextual updates in real time, leading to static outputs that may not align with evolving needs.

The Unified Field of Adaptive Potential (UFAP) provides a foundational construct for rethinking the design of generative AI systems. In this context, UFAP represents the multidimensional space of possible outputs that the model can generate, conditioned on user input, contextual factors, and prior outputs. By conceptualizing the model's output space as a dynamic, probabilistic landscape, UFAP enables generative AI to explore diverse pathways while maintaining coherence.

For example, in a generative text model, UFAP can guide the exploration of potential narrative arcs or stylistic variations, ensuring that the system balances creativity with contextual alignment. This probabilistic framework also facilitates real-time recalibration, where the model adjusts its trajectory based on user feedback or updated contextual information. This dynamic adaptability ensures that generative AI systems remain responsive and relevant in evolving environments.

The SPARC Framework operationalizes adaptability in generative AI by combining the Spectrum of Possibility with Recursive Choice. The Spectrum of Possibility allows the model to maintain diversity in its outputs, ensuring that it explores a wide range of creative or functional options. Recursive Choice introduces iterative refinement processes that incorporate user feedback and contextual updates into the system's decision-making.

For instance:

• Real-Time Refinement: A text generator can revise its ongoing output based on real-time feedback, such as a user clarifying tone, theme, or specific preferences mid-generation.

• Iterative Learning: Recursive feedback enables the system to adapt to new contexts by updating its internal representation or modifying its decision pathways, improving its performance over time without requiring retraining on static datasets.

The integration of Neodynamic principles into generative AI opens new possibilities across a range of applications:

1.   Creative Industries: Adaptive AI systems can generate context-aware content for storytelling, marketing, or design, dynamically adjusting to user preferences or cultural trends. For example, a content generator for video games could align its narratives with player decisions, fostering emergent and personalized experiences.

2.   Healthcare Communication: Generative AI systems in healthcare can produce personalized patient education materials, adapting dynamically to the individual's medical history, literacy level, and real-time feedback from healthcare providers.

3.   Collaborative Tools: AI-powered collaborative platforms, such as coding assistants or design software, can use Neodynamic principles to enhance user productivity by generating outputs that align with evolving project goals and real-time user input.

Consider an adaptive writing assistant that integrates UFAP and SPARC to produce tailored narratives for various use cases. When generating a novel, the system begins by exploring multiple narrative structures (Spectrum of Possibility), balancing creativity with genre conventions. As the author provides feedback—e.g., specifying preferred themes or rejecting certain plot elements—the system dynamically adjusts its output trajectory (Recursive Choice), refining the narrative while maintaining coherence. This adaptive process enables the system to produce high-quality, context-sensitive content that evolves in alignment with user goals.

Neodynamics revolutionizes generative AI design by embedding adaptability and emergent coherence into system architectures. Through constructs like UFAP and SPARC, these systems can overcome the limitations of static optimization, enabling real-time adaptability, personalized outputs, and scalable creativity. This application demonstrates the transformative potential of Neodynamics in rethinking how generative AI operates in dynamic and complex environments.

Adaptive Governance

Adaptive governance, as informed by Neodynamics, provides a framework for designing systems of governance that are flexible, responsive, and capable of evolving in real time. Traditional governance structures often rely on hierarchical decision-making processes and rigid policies, which prioritize stability and predictability over adaptability. While effective in stable contexts, these systems frequently fail to respond adequately to dynamic crises, such as pandemics, climate change, and geopolitical instability. Neodynamics redefines governance by embedding adaptability, feedback integration, and emergent coherence into the core of institutional design.

1.   Inflexibility: Static governance models are ill-equipped to recalibrate policies in response to rapidly changing circumstances. For example, during the COVID-19 pandemic, many governments experienced delays in updating public health measures, resulting in prolonged disruptions and preventable losses (Ostrom, 2009).

2.   Centralized Control: Hierarchical structures concentrate decision-making power at the top, limiting the flow of real-time feedback from local or community-level actors who often have critical insights into emerging issues.

3.   Siloed Operations: Bureaucratic systems often operate in silos, preventing effective communication and coordination across sectors, which is essential in addressing interconnected challenges like climate change or economic crises.

These limitations underscore the need for governance systems that are not only resilient but inherently adaptive. The Unified Field of Adaptive Potential (UFAP) provides a foundation for rethinking policy design and institutional decision-making. In governance, UFAP represents the spectrum of potential policy actions, interventions, and strategies available to address a given issue. By mapping this multidimensional space, policymakers can explore a wide range of adaptive pathways rather than committing prematurely to static solutions.

For instance:

•   In climate policy, UFAP allows governments to evaluate multiple strategies, such as renewable energy adoption, carbon pricing, and community-based adaptation. By treating policy decisions as part of a dynamic landscape, UFAP ensures that governance systems remain flexible and can recalibrate as new data and feedback emerge.

The SPARC Framework transforms UFAP's theoretical insights into actionable processes for adaptive governance. The Spectrum of Possibility enables governments to maintain diverse policy options, ensuring that decisions are not constrained by rigid assumptions. Recursive Choice facilitates iterative policy evaluation and adjustment, allowing governance systems to integrate feedback from stakeholders and real-time data.

Key applications of SPARC in governance include:

1.   Real-Time Policy Adjustments: Policies are continuously evaluated based on their effectiveness and recalibrated as conditions change. For example, during an economic crisis, fiscal policies such as stimulus packages can be adjusted iteratively in response to shifting unemployment rates or market conditions.

2.   Community-Led Feedback Integration: Recursive Choice mechanisms ensure that feedback from local stakeholders is systematically incorporated into decision-making. This decentralization fosters emergent coherence, where governance systems align with diverse societal needs while maintaining overarching objectives.

3.   Scenario-Based Planning: The Spectrum of Possibility allows governments to simulate multiple future scenarios and design adaptive strategies that remain effective across a range of potential outcomes.

Consider a disaster management system designed using Neodynamic principles. At its core, the system incorporates UFAP to explore adaptive responses to natural disasters, such as floods or hurricanes. The Spectrum of Possibility includes actions ranging from evacuation strategies and resource allocation to long-term infrastructure planning. Recursive Choice mechanisms allow the system to adjust dynamically based on real-time data from weather models, community reports, and resource availability.

For example, during a hurricane, the system could recalibrate evacuation routes based on traffic patterns and shelter capacity, ensuring that resources are allocated efficiently while minimizing risks to affected populations. This iterative process not only improves immediate disaster response but also informs long-term planning by integrating lessons learned into future scenarios.

Emergent coherence is critical for aligning diverse stakeholders and objectives within governance systems.

Neodynamics achieves this through feedback-driven interactions that balance local autonomy with global coordination. For instance, decentralized governance models can enable community-level actors to address localized issues while maintaining alignment with national or international priorities. This approach fosters coherence without imposing rigid, top-down control, enabling governance systems to adapt organically to shifting societal needs.

Implementing Neodynamic principles in governance poses challenges, including:

- Data Integration: Adaptive governance requires robust systems for collecting, analyzing, and integrating diverse data streams in real time.

- Balancing Autonomy and Coordination: Ensuring coherence across decentralized actors while maintaining local flexibility is a complex task that requires carefully calibrated feedback mechanisms.

- Ethical Concerns: Iterative policy adjustments must be transparent and inclusive to avoid unintended biases or inequities.

By addressing these challenges, Neodynamics offers a pathway for transforming governance systems into adaptive, feedback-driven entities capable of navigating complexity. Adaptive governance, as informed by Neodynamics, redefines how policies and institutions respond to uncertainty and change. By embedding constructs like UFAP and SPARC into the decision-making process, governance systems become dynamic, resilient, and capable of fostering emergent coherence. This approach ensures that institutions not only survive disruptions but evolve to meet the demands of an increasingly complex and interconnected world.

Healthcare

Healthcare systems face significant challenges in addressing dynamic and individualized needs while managing resource constraints and complex feedback loops. Traditional healthcare models often rely on static protocols and standardized treatments, which can fail to account for the variability in patient responses, evolving medical knowledge, and shifting public health conditions. Neodynamics provides a transformative framework for embedding adaptability, feedback-driven processes, and emergent coherence into healthcare systems, enabling more effective and personalized care.

1.   Standardized Protocols: Medical guidelines often assume homogeneity among patients, applying fixed treatment pathways that may not suit individual variability. For example, patients with the same diagnosis may exhibit vastly different responses to identical treatments due to genetic, environmental, or lifestyle factors (Collins & Varmus, 2015).

2.   Inflexible Infrastructure: Healthcare infrastructure is often designed for predictable demand, making it difficult to respond effectively to surges, such as during pandemics or natural disasters.

3.   Data Silos: Fragmented data systems hinder the flow of critical information between providers, patients, and institutions, limiting the ability to integrate real-time feedback into clinical and administrative decision-making.

These limitations create inefficiencies and inequities, highlighting the need for a more dynamic and adaptive approach to healthcare delivery. The Unified Field of Adaptive Potential (UFAP) offers a conceptual and operational tool for navigating the complexity of healthcare decision-making. In this context, UFAP represents the multidimensional space of possible diagnostic, therapeutic, and administrative actions, incorporating real-time data, patient-specific factors, and evolving medical knowledge.

For example:

• Personalized Treatment Pathways: UFAP enables clinicians to explore a wide range of treatment options tailored to individual patients. By integrating genetic data, biomarkers, and patient feedback, healthcare systems can dynamically identify and refine optimal therapeutic strategies.

• Dynamic Resource Allocation: In public health, UFAP can model potential responses to emerging crises, such as vaccine distribution strategies during an epidemic. By mapping the adaptive landscape of resource deployment, systems can optimize outcomes while maintaining flexibility.

The SPARC Framework operationalizes adaptability in healthcare by combining the Spectrum of Possibility with Recursive Choice. These mechanisms ensure that healthcare systems remain responsive to individual and population-level needs, integrating feedback into every level of decision-making.

Applications include:

• Real-Time Monitoring and Adjustment: Wearable health devices and remote monitoring systems provide continuous feedback on patient health, enabling dynamic adjustments to treatments. For example, a diabetic patient's insulin regimen could be automatically recalibrated based on real-time glucose monitoring.

• Iterative Public Health Strategies: During a public health emergency, SPARC allows policymakers to evaluate and refine interventions such as lockdown measures, vaccination campaigns, or public messaging strategies. This iterative approach ensures that responses evolve in alignment with changing conditions and community feedback.

Emergent coherence is critical for aligning diverse stakeholders and objectives in healthcare, from individual clinicians and patients to hospitals and public health agencies. Neodynamics fosters coherence by embedding feedback-driven processes into healthcare workflows, enabling decentralized actors to coordinate effectively without relying on rigid, top-down control.

For example:

• Decentralized Care Networks: A network of community health centers operating under Neodynamic principles could self-organize to address localized needs, such as managing outbreaks or improving access to preventative care. Feedback mechanisms ensure that these centers align with broader public health objectives while adapting to local contexts.

The implementation of Neodynamics in healthcare faces several challenges:

• Data Integration: Effective use of UFAP and SPARC requires robust systems for collecting, sharing, and analyzing healthcare data across institutions and platforms.

• Ethical Considerations: Dynamic decision-making processes must prioritize patient consent, privacy, and equity to avoid unintended harms or disparities.

• Systemic Resistance: Institutional inertia and regulatory frameworks may slow the adoption of adaptive healthcare models.

Despite these challenges, Neodynamics offers a pathway for overcoming the limitations of static healthcare systems. By embedding adaptability and feedback into every level of care delivery, it enables systems to respond effectively to complexity and change. The application of Neodynamics to healthcare represents a paradigm shift in how care is delivered, managed, and evaluated.

Through constructs like UFAP and SPARC, healthcare systems can achieve greater personalization, resilience, and alignment with patient and public health needs. This approach not only addresses the inefficiencies of traditional models but also provides a framework for navigating the complexity of modern healthcare in a way that empowers individuals and communities alike.

Education

Education systems, like healthcare and governance, are often constrained by static structures that fail to adapt to the dynamic needs of learners, educators, and evolving societal demands. Traditional educational frameworks rely on standardized curricula, rigid assessment models, and one-size-fits-all teaching approaches, which can marginalize diverse learners and fail to prepare students for the complexities of the modern world. Neodynamics offers a transformative approach to education by embedding adaptability, feedback-driven processes, and emergent coherence into learning systems, fostering environments where both students and institutions can thrive under dynamic conditions.

1.   Standardized Curricula: Fixed curricular models prioritize uniformity over adaptability, often failing to address

individual learning needs or diverse cultural contexts. This rigidity results in disengagement and inequity, particularly for students with nontraditional learning styles or socio-economic challenges (Heckman, 2011).

2.   Rigid Assessment Practices: Standardized testing focuses on static benchmarks of performance, neglecting the iterative nature of learning and the broader skills required for adaptability, creativity, and critical thinking.

3.   Inflexible Institutions: Traditional educational institutions often struggle to adapt to rapid changes in technology, labor markets, and societal values, leaving graduates unprepared for emerging challenges and opportunities.

These limitations highlight the need for adaptive educational systems that can respond dynamically to diverse and evolving demands. The Unified Field of Adaptive Potential (UFAP) redefines how educational systems conceptualize and navigate the spectrum of learning pathways available to students. By treating education as a dynamic process of exploration and discovery, UFAP enables systems to move beyond fixed curricular models and embrace diverse learning trajectories.

Applications include:

• Personalized Learning Pathways: UFAP maps the potential learning trajectories for each student based on their strengths, interests, and developmental needs. For example, adaptive learning platforms can use UFAP to guide students through personalized educational content, ensuring alignment with their unique learning styles and goals.

• Dynamic Curricular Design: Educators and institutions can use UFAP to evaluate and update curricula in real time, incorporating new knowledge, technologies, and societal priorities into their teaching frameworks.

The SPARC Framework operationalizes feedback-driven adaptability in education by combining the Spectrum of Possibility with Recursive Choice. These mechanisms ensure that educational systems remain responsive to the needs of learners, educators, and communities, integrating feedback into every level of the learning process.

Key applications include:

1. Iterative Learning Systems: Adaptive learning technologies, guided by SPARC, provide continuous feedback to

students and educators, enabling iterative refinements to instructional approaches and content delivery.

2.   Dynamic Assessment Models: Recursive Choice allows for the development of assessment practices that evolve in alignment with student progress, moving away from static tests toward dynamic, formative evaluations that support growth and understanding.

3.   Institutional Adaptability: Schools and universities can use SPARC to iteratively refine their policies, programs, and resource allocations, ensuring alignment with community needs and global trends.

Emergent coherence is critical for aligning the diverse actors and objectives within educational systems, from students and teachers to policymakers and employers. Neodynamics fosters coherence by embedding feedback-driven processes that allow for decentralized, organic alignment across these stakeholders.

For example:

•   Collaborative Learning Networks: Decentralized networks of educators, students, and institutions can self-organize to address specific learning challenges or opportunities, such as creating interdisciplinary programs or responding to local labor market demands. Feedback mechanisms ensure that these networks remain aligned with broader educational goals while adapting to local contexts.

The application of Neodynamics to education poses several challenges:

- Technological Integration: Adaptive learning systems require robust technological infrastructures to collect, analyze, and act on real-time feedback.

- Equity and Access: Ensuring that all students benefit from adaptive education requires addressing disparities in access to technology, resources, and personalized support.

- Cultural Resistance: Transitioning from traditional educational paradigms to adaptive models may face resistance from educators, institutions, and policymakers accustomed to static systems.

Despite these challenges, the opportunities for transformation are profound. Adaptive educational systems informed by Neodynamics can prepare students not only for current demands but also for navigating the uncertainties of a rapidly changing world.

Neodynamics transforms education by embedding adaptability, feedback, and emergent coherence into learning systems. Through constructs like UFAP and SPARC, these systems can evolve dynamically to meet the diverse needs of learners and institutions. By prioritizing personalization, iterative growth, and decentralized alignment, Neodynamic education fosters environments where all stakeholders can thrive, ensuring relevance and resilience in the face of complexity and change.

Supply Chains

Supply chains are foundational to global commerce, enabling the production, distribution, and consumption of goods and services. However, traditional supply chain models are often brittle, designed for efficiency under stable conditions rather than adaptability in the face of disruptions. Natural disasters, pandemics, geopolitical tensions, and demand fluctuations frequently expose the limitations of these static systems, leading to cascading failures. Neodynamics offers a paradigm shift in supply chain design by embedding adaptability, feedback-driven processes, and emergent coherence, ensuring resilience and efficiency under dynamic conditions.

1.   Optimization for Stability: Traditional supply chains prioritize cost minimization and just-in-time (JIT) inventory management, leaving them vulnerable to unexpected disruptions (Ivanov & Dolgui, 2020).

2.   Lack of Real-Time Feedback: Static models often fail to incorporate real-time data, resulting in delayed responses to demand fluctuations or supply shortages.

3.   Fragmentation: Siloed operations across suppliers, manufacturers, and distributors hinder collaboration and coordination, exacerbating disruptions.

These limitations reveal the need for adaptive supply chain systems capable of navigating uncertainty and maintaining functionality across diverse scenarios. The Unified Field of Adaptive Potential (UFAP) provides a framework for conceptualizing the full range of actions available to a supply chain under changing conditions. By modeling supply chains as dynamic, multidimensional systems, UFAP enables stakeholders to explore adaptive pathways that balance efficiency, resilience, and sustainability.

Applications include:

- Dynamic Inventory Management: UFAP allows supply chains to model optimal inventory strategies under varying conditions, such as balancing JIT principles with strategic stockpiling for critical resources.

- Flexible Logistics Planning: By mapping transportation and distribution options, UFAP supports real-time recalibration of logistics networks in response to disruptions, such as port closures or fuel shortages.

The SPARC Framework operationalizes feedback-driven processes within supply chains, ensuring that decisions are continuously refined based on real-time data and emerging trends. The Spectrum of Possibility allows supply chains to explore diverse scenarios, while Recursive Choice ensures iterative adjustments to strategies.

Key applications include:

1.   Real-Time Demand-Supply Balancing: SPARC enables dynamic matching of supply with fluctuating demand through real-time data from sensors, market analytics, and customer feedback.

2.   Iterative Supplier Selection: Recursive feedback allows supply chains to evaluate supplier performance dynamically, ensuring continuity and quality while adapting to shifting conditions.

3.   Responsive Production Systems: Factories and manufacturers can iteratively adjust production schedules and processes based on raw material availability and demand signals.

Emergent coherence is essential for aligning the diverse actors and objectives within supply chains, from raw material suppliers to end consumers. Neodynamics fosters this coherence by embedding decentralized feedback loops that allow for organic coordination across the network.

For example:

•   Collaborative Ecosystems: A Neodynamic supply chain might consist of decentralized, interdependent nodes that self-organize to address disruptions. Local suppliers, distributors, and logistics providers can autonomously adjust their operations while maintaining alignment with the overarching objectives of the supply network.

This emergent alignment enables supply chains to absorb shocks, reorganize dynamically, and maintain functionality even in the face of severe disruptions.

Implementing Neodynamic principles in supply chains presents challenges:

• Data Integration: Adaptive supply chains require robust systems for collecting, sharing, and analyzing data across all nodes in the network.

• Balancing Efficiency and Resilience: Overemphasizing adaptability can lead to inefficiencies, while excessive optimization for stability reduces flexibility.

• Cultural Shifts: Transitioning to decentralized, feedback-driven supply networks requires changes in mindset and practices across organizations.

Despite these challenges, Neodynamics offers transformative opportunities for supply chain design. By embedding adaptability and feedback at every level, supply chains can become more resilient, sustainable, and responsive to the complexities of a globalized economy. Neodynamics redefines supply chain management by integrating adaptability, feedback, and emergent coherence into system design. Through constructs like UFAP and SPARC, supply chains can dynamically navigate disruptions, align diverse stakeholders, and optimize performance under uncertainty. This approach ensures that supply chains remain resilient and efficient, meeting the demands of an increasingly complex and interconnected world.

Robotics

Robotics, as a field, is rapidly evolving to address challenges across industrial automation, healthcare, disaster response, and everyday applications. However, many robotic systems remain constrained by static designs and pre-programmed behaviors, limiting their ability to adapt to dynamic environments or unforeseen challenges. Neodynamics offers a framework for embedding adaptability, feedback-driven processes, and emergent coherence into robotic systems, enabling them to operate autonomously and flexibly in complex, real-world conditions.

1.   Static Programming: Many robots rely on pre-defined algorithms that perform well under expected conditions but struggle with variability or novel scenarios. For instance, industrial robots are

often optimized for repetitive tasks but lack the flexibility to handle irregularities in materials or workflows.

2.   Centralized Control: Traditional robotic systems depend heavily on centralized processing, which limits their scalability and responsiveness in decentralized environments.

3.   Limited Collaboration: Robotic systems often operate in isolation, lacking the ability to coordinate with other robots or human agents in real time.

These limitations highlight the need for adaptable robotic systems that can learn, evolve, and collaborate autonomously. The Unified Field of Adaptive Potential (UFAP) enables robotics to move beyond static programming by modeling the multidimensional space of possible actions a robot can take in response to environmental inputs. By treating adaptability as a dynamic process of exploration and optimization, UFAP allows robotic systems to operate effectively in unpredictable conditions.

Applications include:

•   Dynamic Motion Planning: UFAP enables robots to continuously evaluate and adjust their trajectories based on obstacles, changing environments, or real-time feedback. For example, an autonomous drone navigating a disaster zone can explore multiple pathways to identify the safest and most efficient route.

•   Adaptive Task Allocation: In multi-robot systems, UFAP can model the potential roles each robot can assume, dynamically reallocating tasks based on performance, resource availability, and emergent priorities.

The SPARC Framework provides the operational tools for embedding adaptability into robotic systems. The Spectrum of Possibility ensures that robots maintain a diverse set of behavioral options, while Recursive Choice facilitates real-time learning and decision-making based on feedback.

Key applications include:

1.   Real-Time Behavioral Adjustment: Robots guided by SPARC can iteratively refine their actions based on environmental feedback. For example, a robotic arm assembling products can adjust its grip strength or positioning in response to sensor data.

2.   Collaborative Swarm Robotics: Recursive feedback allows swarms of robots to self-organize and coordinate their actions dynamically. For instance, a group of robotic search-and-rescue

agents can divide tasks autonomously, adaptively focusing resources on areas with the greatest need.

3.   Human-Robot Interaction: SPARC enhances human-robot collaboration by enabling robots to adapt their behaviors to user preferences, gestures, or verbal commands in real time, fostering seamless and intuitive interactions.

Emergent coherence is critical for ensuring that robots operating in decentralized environments align their actions with broader system objectives. Neodynamics fosters this coherence by embedding feedback-driven processes that enable robots to collaborate effectively with each other and with human agents.

For example:

•   Decentralized Warehouse Robotics: In a Neodynamic warehouse, autonomous robots can self-organize to optimize inventory management. Using feedback from real-time demand data and peer interactions, the robots can dynamically adjust their roles, such as retrieving items, restocking shelves, or reorganizing storage layouts.

This emergent alignment ensures that robotic systems remain flexible and responsive while maintaining coherence with overarching goals.

Applying Neodynamics to robotics involves several challenges:

• Computational Complexity: Modeling the Spectrum of Possibility and integrating real-time feedback require advanced computational tools and architectures.

• Safety and Reliability: Feedback-driven systems must ensure stability and safety, particularly in applications like healthcare or autonomous driving, where errors can have critical consequences.

• Ethical Considerations: Robots operating autonomously in human environments must adhere to ethical principles, such as fairness, accountability, and transparency.

Despite these challenges, Neodynamics unlocks transformative opportunities in robotics. By embedding adaptability, feedback, and emergent coherence into system design, robots can achieve higher levels of autonomy, flexibility, and collaboration, enabling them to address complex, real-world challenges. Neodynamics reimagines robotics by introducing principles of adaptability, feedback-driven learning, and emergent coherence. Through constructs like UFAP and SPARC, robotic systems can move beyond static programming and centralized control, achieving autonomy and responsiveness in dynamic environments. This approach ensures that robotics not only meets the demands of today's applications but also evolves to address the uncertainties and complexities of the future.

Quantum Computing

Quantum computing, with its potential to solve problems far beyond the reach of classical computers, represents one of the most transformative technological frontiers. However, the design and operation of quantum systems remain constrained by static architectures, limited adaptability, and fragile coherence in the presence of environmental noise. Neodynamics offers a framework for embedding adaptability and feedback-driven processes into quantum computing systems, addressing these challenges and enabling quantum technologies to thrive in complex, dynamic environments.

1.   Fragile Quantum States: Quantum systems are highly sensitive to environmental noise and decoherence, which disrupt

their functionality. Current error correction methods are computationally intensive and often insufficient for large-scale quantum systems (Shor, 1995).

2.   Static Algorithms: Quantum algorithms, such as Shor's factoring algorithm or Grover's search algorithm, are designed for specific problems and lack flexibility to adapt dynamically to variations in input or problem structure.

3.   Resource Constraints: The physical and computational resources required for quantum computing are immense, creating bottlenecks in scalability and performance.

These limitations underscore the need for adaptable quantum systems capable of dynamically recalibrating their operations in response to environmental and computational challenges. The Unified Field of Adaptive Potential (UFAP) offers a conceptual and operational framework for modeling the adaptive landscape of quantum systems. In this context, UFAP represents the multidimensional space of quantum states, configurations, and operational pathways available to a quantum computer under varying conditions.

Applications include:

• Dynamic Quantum Error Correction: UFAP enables quantum systems to explore a range of error correction strategies dynamically, optimizing their approach based on real-time feedback about environmental noise and qubit stability.

• Adaptive Quantum State Management: By mapping the potential configurations of quantum states, UFAP allows systems to dynamically reconfigure qubits to optimize computational efficiency and coherence.

The SPARC Framework operationalizes adaptability in quantum systems by embedding the Spectrum of Possibility and Recursive Choice into quantum algorithms and architectures. These mechanisms ensure that quantum systems can explore diverse pathways and iteratively refine their operations based on feedback. Key applications include:

1. Dynamic Quantum Algorithms: SPARC enables quantum systems to adapt their algorithms in real time, optimizing operations for variable inputs or problem structures. For instance, a quantum optimization algorithm could recalibrate its approach based on evolving constraints or objectives.

2. Iterative Feedback for Quantum Simulations: Recursive feedback allows quantum systems to refine their simulations dynamically, improving accuracy and relevance as new data or conditions emerge.

3. Resource Allocation in Quantum Networks: In distributed quantum computing environments, SPARC ensures that computational and communication resources are dynamically allocated based on real-time demand and system performance.

Emergent coherence is particularly relevant in quantum computing, where the alignment of quantum states across a system is critical for functionality. Neodynamics fosters this coherence by embedding feedback-driven processes that enable quantum systems to self-organize and maintain alignment in the presence of noise and other disruptions.

For example:

• Decentralized Quantum Networks: In a quantum communication network, Neodynamics enables nodes to self-organize, dynamically recalibrating their operations to maximize coherence and minimize error rates. Feedback from the network ensures that quantum states remain aligned, even under varying conditions.

This emergent alignment ensures that quantum systems maintain functionality and efficiency while adapting to the complexities of real-world applications.

Implementing Neodynamic principles in quantum computing poses unique challenges:

• Noise and Decoherence: Quantum systems require robust mechanisms for managing environmental noise and maintaining coherence across qubits.

• Computational Complexity: Modeling the adaptive landscape of quantum systems and integrating feedback-driven processes requires significant computational and architectural advancements.

• Scalability: As quantum systems grow in scale, ensuring alignment and coherence across distributed qubits becomes increasingly challenging.

Despite these challenges, Neodynamics offers transformative opportunities for quantum computing. By embedding adaptability and feedback into quantum algorithms and architectures, quantum systems can achieve greater robustness, scalability, and functionality, unlocking new possibilities in optimization, cryptography, and simulation. Neodynamics transforms quantum computing by introducing principles of adaptability, feedback, and emergent coherence into quantum system design. Through constructs like UFAP and SPARC, quantum computers can move beyond static architectures and fragile coherence, achieving the flexibility and resilience needed to thrive in complex environments. This approach ensures that quantum technologies not only overcome current limitations but also evolve to meet the demands of future applications.

Climate Change and Ecological Systems

Climate change and the degradation of ecological systems represent some of the most complex and pressing challenges of the 21st century. These challenges are characterized by their dynamic, nonlinear nature, with feedback loops and cascading effects that span across scales and domains. Traditional approaches to addressing climate and ecological issues often rely on rigid policies, centralized control, and static modeling, which fail to capture the complexity and adaptability required for effective action. Neodynamics provides a transformative framework for embedding adaptability, feedback-driven processes, and emergent coherence into strategies for mitigating climate change and restoring ecological systems.

1.   Predictive Failures: Static models often fail to account for nonlinear feedback loops, such as tipping points in ecosystems or rapid shifts in climate dynamics. For example, the accelerated melting of polar ice sheets triggers feedback that amplifies global warming (Lenton et al., 2008).

2.   Top-Down Governance: Centralized approaches to climate policy often ignore the localized dynamics of ecosystems and communities, leading to inefficient or inequitable solutions.

3.   Inflexibility in Resource Management: Rigid management strategies, such as fixed quotas for water use or fishing, often fail to adapt to fluctuations in environmental conditions, resulting in resource depletion or ecosystem collapse.

These limitations highlight the need for adaptive frameworks that can navigate the uncertainty and complexity inherent in climate and ecological systems. The Unified Field of Adaptive Potential (UFAP) provides a powerful tool for modeling the adaptive pathways available for mitigating climate change and managing ecosystems. UFAP conceptualizes the multidimensional landscape of possible actions, allowing policymakers, scientists, and communities to explore diverse strategies and dynamically recalibrate their approaches.

Applications include:

•   Adaptive Climate Policy Design: UFAP can guide the exploration of policy actions such as carbon pricing, renewable energy deployment, and reforestation efforts, dynamically adjusting strategies based on feedback from economic, environmental, and social systems.

•   Ecosystem Restoration Planning: UFAP allows conservationists to model the potential outcomes of restoration efforts, such as reintroducing species, rebuilding habitats, or altering land use, ensuring that interventions remain flexible and responsive to changing ecological conditions.

The SPARC Framework operationalizes adaptability in climate and ecological systems by embedding the Spectrum of Possibility and Recursive Choice into decision-making processes.

These mechanisms ensure that strategies evolve iteratively based on real-time data and emergent conditions.

Key applications include:

1.   Dynamic Resource Management: SPARC enables real-time adjustments to resource management practices, such as water allocation or agricultural planning, based on weather patterns, ecosystem health, and community needs.

2.   Iterative Carbon Offset Strategies: Recursive feedback allows policymakers to evaluate the effectiveness of carbon offset initiatives, such as afforestation or renewable energy investments, and refine their approaches as new data emerges.

3.   Localized Climate Adaptation: The Spectrum of Possibility ensures that communities can explore diverse adaptation pathways, from building resilient infrastructure to altering agricultural practices, while maintaining alignment with broader climate goals.

Emergent coherence is essential for aligning the diverse actors and objectives involved in climate and ecological management, from governments and NGOs to local communities and ecosystems themselves. Neodynamics fosters this coherence by embedding decentralized feedback loops that enable dynamic alignment across scales.

For example:

• Decentralized Conservation Networks: A Neodynamic framework for conservation might involve a network of local initiatives that self-organize to protect biodiversity. By integrating feedback from local ecosystems, these initiatives can dynamically adjust their strategies while contributing to regional and global conservation goals.

Implementing Neodynamics in climate and ecological systems presents challenges:

• Complexity of Feedback Loops: Modeling and managing the intricate feedback loops in climate and ecological systems requires advanced computational tools and interdisciplinary collaboration.

• Equity and Inclusivity: Adaptive strategies must prioritize equity, ensuring that marginalized communities are not disproportionately burdened by climate policies or ecological interventions.

- Scalability: Coordinating adaptive actions across scales, from local communities to international coalitions, requires robust mechanisms for communication and alignment.

Despite these challenges, Neodynamics offers transformative opportunities for addressing climate change and ecological degradation. By embedding adaptability, feedback, and emergent coherence into strategies and policies, Neodynamics enables dynamic and effective responses to the complexity of these challenges. Neodynamics redefines approaches to climate change and ecological systems by introducing principles of adaptability, feedback-driven processes, and emergent coherence. Through constructs like UFAP and SPARC, this framework provides actionable tools for designing policies, managing ecosystems, and fostering alignment among diverse stakeholders. By embracing the complexity of climate and ecological systems, Neodynamics offers a pathway for addressing these global challenges with resilience, innovation, and equity.

Urban Planning and Smart Cities

Urban areas are increasingly dynamic systems characterized by dense populations, complex infrastructures, and intersecting challenges such as climate adaptation, housing shortages, and traffic congestion. Traditional urban planning models rely on static, long-term master plans that fail to adapt to real-time demands or unexpected disruptions. Neodynamics offers a transformative framework for designing adaptive, feedback-driven, and resilient urban systems capable of thriving in the face of rapid change.

1.   Inflexibility: Static master plans often fail to account for evolving demographics, economic trends, or environmental risks, leaving cities unprepared for crises such as extreme weather or rapid migration.

2.   Siloed Systems: Infrastructure, transportation, energy, and housing systems are often managed independently, reducing coordination and creating inefficiencies.

3.   Inequity: Static models frequently fail to address the diverse and evolving needs of marginalized communities, exacerbating social inequalities.

These limitations underscore the need for adaptive frameworks that integrate real-time data, community input, and system-level coordination. The Unified Field of Adaptive Potential (UFAP) provides a framework for exploring dynamic pathways in urban planning. UFAP models the spectrum of adaptive actions available to urban systems, integrating data from IoT sensors, environmental models, and public engagement platforms.

Applications include:

•   Dynamic Transportation Networks: UFAP can map and optimize adaptive transit systems that respond in real time to changes in demand, weather conditions, or infrastructure disruptions.

•   Resilient Energy Grids: By modeling the adaptive potential of distributed energy systems, UFAP supports dynamic allocation of resources, integrating renewable energy sources and storage technologies.

•   Smart Housing Solutions: UFAP enables cities to explore adaptive housing policies and modular infrastructure that align with changing population needs.

The SPARC Framework ensures that urban systems continuously evolve through iterative decision-making and feedback-driven processes. The Spectrum of Possibility allows cities to maintain a range of adaptive policies and infrastructure pathways, while Recursive Choice integrates real-time data into planning and operational adjustments.

Key applications include:

1.   Iterative Urban Design: Recursive feedback loops enable urban planners to refine zoning policies, public space design, and development priorities dynamically.

2.   Community Engagement Platforms: The SPARC Framework supports decentralized feedback systems where residents contribute real-time data and insights, enabling cities to adapt policies to reflect local needs.

3.   Disaster Preparedness and Response: SPARC ensures that cities dynamically recalibrate evacuation plans, resource distribution, and recovery strategies in response to real-time information during emergencies.

Neodynamics fosters emergent coherence in urban systems by aligning diverse actors and objectives across infrastructure, governance, and community networks. Feedback-driven processes ensure that decentralized components, such as local energy grids or neighborhood councils, self-organize into functional and resilient wholes.

For example:

- Integrated Mobility Systems: In a Neodynamic city, public transit, ride-sharing services, and autonomous vehicles could self-coordinate through feedback loops to optimize efficiency and reduce congestion, aligning with broader sustainability goals.

By embedding adaptability, feedback, and emergent coherence into urban planning, Neodynamics transforms cities into dynamic systems capable of thriving in complexity. This approach ensures that urban areas remain resilient, equitable, and innovative, addressing the challenges of the 21st century with agility and vision.

Financial Systems and Economic Policy

Financial systems and economic policies operate within inherently complex environments, marked by interdependencies, market volatility, and unpredictable shocks. Static models often fail to account for the nonlinear dynamics of these systems, leaving them vulnerable to crises and inefficiencies. Neodynamics provides a transformative approach to designing adaptive financial systems and policymaking frameworks, embedding real-time feedback, iterative decision-making, and emergent coherence to ensure stability and innovation.

1.   Predictive Inadequacies: Traditional economic models rely on equilibrium assumptions and historical data, making them ill-equipped to respond to sudden disruptions, such as financial crises or global trade shocks.

2.   Delayed Policy Adjustments: Economic policies often lag behind real-time developments, relying on periodic reviews rather than continuous recalibration.

3.   Fragmented Regulation: Financial systems, composed of diverse institutions and markets, are governed by siloed regulatory frameworks that hinder systemic coordination and adaptability.

These limitations underscore the need for a more dynamic approach to economic and financial systems design. The Unified Field of Adaptive Potential (UFAP) provides a framework for exploring the adaptive landscape of financial systems and policies. By modeling the multidimensional space of possible actions, UFAP allows stakeholders to evaluate and implement dynamic strategies for resilience and growth.

Applications include:

•   Dynamic Monetary Policy: Central banks can use UFAP to simulate and implement real-time adjustments to interest rates, quantitative easing, and liquidity measures in response to market conditions.

• Risk Management and Contingency Planning: UFAP enables financial institutions to explore diverse risk mitigation strategies, ensuring resilience to unexpected shocks, such as market crashes or geopolitical events.

• Adaptive Fiscal Policies: Governments can model resource allocation and taxation strategies within UFAP to balance economic growth, equity, and sustainability dynamically.

The SPARC Framework operationalizes adaptability in financial systems through the Spectrum of Possibility and Recursive Choice. These mechanisms enable continuous evaluation and iterative refinement of policies and strategies, ensuring responsiveness to emerging trends.

Key applications include:

1. Iterative Crisis Management: During financial crises, Recursive Choice enables policymakers to refine interventions— such as bailouts, stimulus measures, or debt restructuring—based on real-time feedback.

2.   Real-Time Market Regulation: SPARC facilitates the continuous adjustment of regulatory measures, such as capital requirements, risk thresholds, or trading rules, to mitigate systemic vulnerabilities.

3.   Dynamic Investment Strategies: Financial institutions can use SPARC to develop adaptive portfolios that align with shifting market conditions and investor priorities.

Emergent coherence ensures that the diverse actors and objectives within financial systems dynamically align, fostering systemic stability and resilience. Neodynamics enables this coherence through decentralized feedback loops that integrate market behavior, regulatory responses, and institutional strategies. For example:

•   Global Financial Networks: A Neodynamic framework could enable global financial institutions and central banks to self-organize in response to currency fluctuations or trade imbalances, dynamically reallocating resources to stabilize markets and support equitable growth.

Applying Neodynamics to financial systems involves several challenges:

- Data Integration: Adaptive financial systems require comprehensive, real-time data collection and analysis across markets, institutions, and geopolitical contexts.

- Coordination Across Scales: Aligning local, national, and global financial strategies requires robust frameworks for communication and collaboration.

- Ethical Considerations: Dynamic policies must be designed to promote equity and prevent unintended consequences, such as exacerbating wealth disparities.

Despite these challenges, Neodynamics offers transformative opportunities for financial systems and economic policy. By embedding adaptability, feedback, and emergent coherence, it enables systems to navigate uncertainty, foster innovation, and achieve long-term stability. Neodynamics reimagines financial systems and economic policymaking as adaptive, feedback-driven entities capable of thriving in complexity. Through constructs like UFAP and SPARC, this framework ensures that economies remain resilient, equitable, and aligned with societal goals, addressing the challenges of a volatile and interconnected global economy.

Social Systems and Crisis Management

Social systems are inherently complex, comprising diverse actors, institutions, and communities interconnected across scales. The rapid pace of modern crises—ranging from natural disasters to pandemics and geopolitical conflicts—exposes the limitations of static social systems and traditional crisis management approaches. Neodynamics provides a transformative framework for embedding adaptability, feedback, and emergent coherence into social systems, ensuring timely, effective, and equitable responses to crises.

1.   Rigid Crisis Protocols: Predefined plans often fail to adapt to the dynamic and unpredictable nature of crises, leading to inefficiencies and missed opportunities for intervention.

2.   Centralized Decision-Making: Hierarchical structures delay responses by limiting the flow of real-time information and disempowering local actors.

3.   Fragmentation: Poor coordination among governments, NGOs, and community stakeholders results in redundant or conflicting actions during crises.

These challenges highlight the need for adaptive social systems that leverage real-time feedback and decentralized collaboration. The Unified Field of Adaptive Potential (UFAP) offers a framework for modeling the adaptive pathways available to social systems during crises. By mapping the spectrum of potential actions, UFAP enables stakeholders to identify dynamic strategies that maximize impact while minimizing disruption.

Applications include:

• Dynamic Resource Allocation: UFAP allows policymakers and humanitarian organizations to explore and implement adaptive strategies for distributing food, water, medical supplies, and shelter during emergencies.

• Community-Led Crisis Response: UFAP facilitates the integration of local knowledge and feedback into broader crisis management frameworks, empowering communities to adaptively address their unique challenges.

• Iterative Recovery Planning: UFAP enables post-crisis rebuilding efforts to evolve dynamically, incorporating real-time data and stakeholder input to ensure long-term resilience.

The SPARC Framework operationalizes adaptability in crisis management by embedding the Spectrum of Possibility and Recursive Choice into decision-making processes. These mechanisms ensure that responses are continuously refined based on real-time feedback and emergent conditions.

Key applications include:

1. Real-Time Crisis Response: Recursive Choice enables governments and NGOs to iteratively adjust evacuation plans, relief distribution strategies, and public health interventions as new information becomes available.

2. Decentralized Coordination Platforms: The SPARC Framework fosters collaboration among diverse actors, enabling self-organized responses that align with overarching crisis objectives.

3. Dynamic Policy Adjustment: Social policies—such as emergency housing or unemployment support—can be continuously evaluated and recalibrated to reflect changing needs during and after a crisis.

Emergent coherence ensures that diverse actors within social systems align dynamically to achieve shared goals. Neodynamics fosters this coherence through feedback loops that integrate local, regional, and global efforts into cohesive, adaptive responses.

For example:

• Collaborative Humanitarian Networks: A Neodynamic framework enables NGOs, governments, and local communities to self-organize in real time, dynamically reallocating resources and responsibilities based on immediate needs and constraints.

This emergent alignment reduces inefficiencies, prevents duplication of effort, and ensures that resources are directed to where they are needed most.

While implementing Neodynamics in social systems presents challenges, it also creates opportunities:

• Data Integration: Effective adaptive systems require robust mechanisms for collecting and analyzing data from diverse sources, including sensors, surveys, and community feedback.

• Equity and Inclusivity: Crisis responses must prioritize the needs of marginalized communities, ensuring that adaptive frameworks promote justice and fairness.

• Systemic Complexity: Managing feedback loops and coordinating decentralized actions across scales demands sophisticated tools and governance structures.

Despite these challenges, Neodynamics offers a pathway for transforming crisis management into a dynamic, responsive process that leverages complexity to drive innovation and resilience. Neodynamics redefines social systems and crisis management as adaptive, feedback-driven networks capable of thriving in uncertainty. Through constructs like UFAP and SPARC, this framework empowers stakeholders to respond dynamically to crises, fostering resilience, equity, and coherence in even the most complex and volatile conditions.

Resistance Movements and Anticolonial Struggles

Resistance movements and anticolonial struggles operate within complex systems of oppression, marked by asymmetrical power dynamics, entrenched hierarchies, and systemic inequities. Traditional approaches to organizing resistance often rely on static strategies or charismatic leadership, which can falter in the face of evolving challenges or co-optation. Neodynamics offers a transformative framework for embedding adaptability, feedback-driven processes, and emergent coherence into resistance movements, empowering them to remain resilient and effective in the pursuit of justice and self-determination.

1.   Over-Reliance on Centralized Leadership: Charismatic leadership can unify movements but also creates vulnerabilities

when leaders are removed or discredited, leaving movements disoriented and fragmented.

2.   Static Tactics: Traditional resistance models often rely on fixed tactics, such as protests or boycotts, which can lose effectiveness over time as oppressive systems adapt to counter them.

3.   Fragmentation: Resistance movements frequently struggle to align diverse groups and objectives, leading to internal divisions and diluted impact.

These limitations emphasize the need for adaptive frameworks that can navigate the complexities of resistance and oppression. The Unified Field of Adaptive Potential (UFAP) models the spectrum of possible actions, strategies, and pathways available to resistance movements in their pursuit of liberation. By conceptualizing resistance as a dynamic, adaptive process, UFAP enables movements to explore diverse tactics and recalibrate in response to changing conditions.

Applications include:

•   Dynamic Strategy Development: UFAP helps movements identify and adapt tactics, such as grassroots organizing, digital activism, and economic disruptions, based on the evolving actions of oppressive systems.

• Resilient Community Networks: By mapping the potential configurations of decentralized community structures, UFAP supports the development of adaptive systems that resist co-optation and sustain long-term engagement.

• Iterative Alliance Building: UFAP enables movements to explore and strengthen coalitions across diverse social, cultural, and political groups, ensuring solidarity while respecting differences.

The SPARC Framework operationalizes adaptability in resistance movements, embedding the Spectrum of Possibility and Recursive Choice into organizing and decision-making processes. These mechanisms ensure that strategies are continuously refined based on real-time feedback and emergent conditions.

Key applications include:

1. Real-Time Tactical Adjustments: Recursive Choice allows movements to refine their tactics dynamically, such as shifting from street protests to digital campaigns in response to surveillance or repression.

2.   Decentralized Organizing Models: SPARC fosters decentralized, leaderless structures that enable local actions to align with broader goals without relying on centralized control.

3.   Iterative Narrative Development: Feedback-driven processes enable movements to adapt their messaging and storytelling to resonate with diverse audiences and counter disinformation campaigns.

Emergent coherence is critical for aligning the diverse actors and objectives within resistance movements, from grassroots organizers to international allies. Neodynamics fosters this coherence by embedding feedback-driven processes that allow movements to self-organize into functional and resilient networks. For example:

• Decentralized Resistance Networks: A Neodynamic approach to resistance might involve decentralized hubs of activists and communities that self-coordinate to challenge systemic oppression. Feedback loops ensure that local actions align with global goals while retaining autonomy and contextual relevance.

This emergent alignment allows movements to adapt to repression, maintain solidarity, and sustain momentum over time.

Implementing Neodynamic principles in resistance and anticolonial struggles presents challenges:

• Coordination Across Scales: Aligning local, regional, and global resistance efforts requires robust frameworks for communication and collaboration.

• Ethical Integrity: Adaptive resistance models must prioritize accountability, ensuring that actions align with values of justice and inclusivity.

• Resource Constraints: Resistance movements often operate with limited resources, requiring innovative strategies to maximize impact while minimizing risk.

Despite these challenges, Neodynamics offers transformative opportunities for resistance and anticolonial movements. By embedding adaptability, feedback, and emergent coherence into organizing frameworks, it empowers movements to navigate complexity, resist co-optation, and sustain long-term struggles for liberation.

Neodynamics redefines resistance and anticolonial struggles as adaptive, feedback-driven networks capable of thriving in uncertainty. Through constructs like UFAP and SPARC, this framework ensures that movements remain resilient, innovative, and aligned with their goals, offering a pathway for justice and self-determination in the face of systemic oppression.

Artificial Life and Complex Simulations

Artificial life (A-life) and complex simulations are at the forefront of understanding emergent behaviors, adaptation, and self-organization in dynamic systems. These fields explore how artificial systems, modeled on biological, ecological, and social processes, can simulate life-like phenomena or evolve in response to changing conditions. Traditional approaches to A-life and simulations often rely on fixed rules or static models, limiting their ability to capture the richness of real-world complexity. Neodynamics provides a framework for embedding adaptability, feedback-driven processes, and emergent coherence into artificial life systems, enabling more robust and realistic simulations.

1.   Rigid Rule Sets: Many A-life systems operate on fixed rules, which constrain their ability to evolve or adapt to new inputs dynamically.

2.   Limited Feedback Integration: Static simulations often lack mechanisms for incorporating real-time feedback, reducing their ability to model dynamic and unpredictable processes.

3.   Predictable Outcomes: Fixed initial conditions and parameters lead to simulations that often produce repetitive or deterministic results, failing to capture the stochastic nature of real-world systems.

These constraints emphasize the need for adaptive frameworks that can capture the complexity and dynamism of life-like processes. The Unified Field of Adaptive Potential (UFAP) provides a framework for modeling the adaptive pathways available to A-life systems and simulations. By conceptualizing adaptability as a probabilistic landscape, UFAP enables these systems to explore diverse evolutionary trajectories and respond dynamically to emergent conditions.

Applications include:

•   Adaptive Rule Evolution: UFAP allows A-life systems to evolve their underlying rules dynamically, simulating processes such as natural selection, mutation, and recombination in more realistic ways.

• Dynamic Environmental Modeling: In ecological simulations, UFAP can model the adaptive responses of species or ecosystems to changing environmental conditions, such as climate shifts or resource scarcity.

• Iterative Scenario Exploration: UFAP supports the development of simulations that iteratively explore a range of scenarios, refining outcomes based on real-time feedback or new data inputs.

The SPARC Framework operationalizes adaptability in A-life systems and simulations by embedding the Spectrum of Possibility and Recursive Choice into their design. These mechanisms ensure that systems evolve iteratively and respond to dynamic inputs.

Key applications include:

1. Real-Time Adaptation: Recursive feedback allows A-life systems to adjust their behaviors or structures dynamically in response to environmental changes or user inputs.

2. Self-Organizing Simulations: SPARC enables simulations to model self-organizing phenomena, such as the

emergence of social hierarchies, predator-prey dynamics, or cooperative behaviors in complex systems.

3.   Exploratory Learning Models: By incorporating the Spectrum of Possibility, simulations can explore diverse potential outcomes, fostering deeper insights into system-level behaviors.

Emergent coherence is essential for ensuring that A-life systems and simulations produce meaningful, life-like behaviors without relying on centralized control or predefined outcomes. Neodynamics fosters this coherence by embedding decentralized feedback loops that allow individual components to self-organize into functional wholes.

For example:

•   Simulated Ecosystems: A Neodynamic approach to ecosystem simulation could involve species and environmental factors interacting through adaptive feedback loops, resulting in emergent behaviors such as population cycles, resource competition, and niche formation.

This emergent alignment ensures that simulations remain robust, realistic, and responsive to both internal dynamics and external perturbations.

Implementing Neodynamics in A-life and simulations presents challenges:

• Computational Demands: Modeling adaptive systems with high-dimensional feedback and probabilistic landscapes requires significant computational power and optimization techniques.

• Validation of Results: Ensuring that simulations accurately reflect real-world phenomena while remaining adaptable and non-deterministic requires careful calibration and validation.

• Interdisciplinary Integration: Effective use of Neodynamics in A-life systems demands collaboration across fields, including biology, physics, computer science, and social sciences.

Despite these challenges, Neodynamics offers transformative opportunities for artificial life and complex simulations. By embedding adaptability, feedback, and emergent coherence, it enables systems to evolve in lifelike ways, fostering deeper understanding and innovation across diverse fields. Neodynamics redefines artificial life and complex simulations as adaptive, feedback-driven systems capable of exploring and modeling the richness of real-world complexity. Through constructs like UFAP and SPARC, this framework ensures that simulations evolve dynamically, offering powerful tools for studying life-like behaviors, emergent phenomena, and systemic resilience.

Cultural Systems and Creative Industries

Cultural systems and creative industries operate in dynamic environments shaped by shifting societal values, technological advancements, and global interconnectedness. Traditional approaches to cultural production and distribution often rely on static frameworks, such as predefined artistic conventions or rigid market structures, which fail to adapt to rapidly changing consumer preferences and cultural contexts. Neodynamics offers a transformative approach for embedding adaptability, feedback, and emergent coherence into cultural systems and creative industries, ensuring sustained relevance and innovation.

1.   Inflexibility to Trends: Static production and distribution models often struggle to respond to evolving consumer tastes or societal shifts, leading to obsolescence or cultural stagnation.

2.   Centralized Gatekeeping: Traditional cultural systems rely on centralized institutions—such as publishers, record labels, or

galleries—that limit diversity and innovation by controlling access to production and distribution.

3.   Lack of Feedback Integration: Static systems often fail to incorporate real-time audience feedback, reducing their ability to refine outputs or align with emerging cultural contexts.

These limitations highlight the need for adaptive frameworks that foster creativity, inclusivity, and responsiveness. The Unified Field of Adaptive Potential (UFAP) provides a conceptual framework for exploring the adaptive possibilities within cultural systems. By modeling the multidimensional space of creative pathways and audience engagement strategies, UFAP enables cultural producers and institutions to dynamically recalibrate their approaches.

Applications include:

•   Adaptive Content Creation: UFAP allows creative industries to explore diverse content production strategies, dynamically adjusting to cultural trends and audience feedback.

•   Dynamic Distribution Models: UFAP supports the development of adaptive distribution channels that leverage digital platforms, streaming services, and decentralized networks to reach diverse audiences.

• Iterative Artistic Development: Artists and cultural institutions can use UFAP to refine their creative outputs dynamically, aligning with societal shifts or global conversations.

The SPARC Framework operationalizes adaptability in creative industries by embedding the Spectrum of Possibility and Recursive Choice into artistic and production processes. These mechanisms ensure that cultural systems continuously evolve and refine their outputs in response to feedback and emergent trends. Key applications include:

1. Real-Time Audience Engagement: Recursive feedback loops enable creators to adjust their work dynamically based on audience responses, fostering greater connection and relevance.

2. Decentralized Content Ecosystems: SPARC supports decentralized networks of creators and producers, enabling adaptive collaborations that align with cultural and market shifts.

3. Iterative Marketing Strategies: Feedback-driven processes allow cultural industries to refine promotional campaigns

dynamically, ensuring alignment with audience sentiment and emerging cultural narratives.

Emergent coherence ensures that diverse actors within cultural systems align dynamically to create shared meaning and collective impact. Neodynamics fosters this coherence by embedding decentralized feedback loops that allow cultural producers, audiences, and institutions to self-organize into cohesive ecosystems.

For example:

• Collaborative Cultural Movements: A Neodynamic framework could enable grassroots cultural movements to self-coordinate, dynamically adapting their narratives and strategies to resonate with local and global audiences while maintaining alignment with overarching goals.

This emergent alignment fosters innovation, inclusivity, and resilience in cultural systems.

Implementing Neodynamics in cultural systems and creative industries presents challenges:

- Cultural Fragmentation: Adaptive systems must balance diversity with coherence, ensuring that innovation does not lead to fragmentation or loss of shared meaning.

- Data-Driven Bias: Feedback mechanisms must be designed to avoid reinforcing biases or privileging dominant cultural narratives over marginalized voices.

- Sustainability: Ensuring that adaptive cultural systems remain economically and ecologically sustainable requires careful resource management and long-term planning.

Despite these challenges, Neodynamics offers transformative opportunities for cultural systems. By embedding adaptability, feedback, and emergent coherence into their design, these systems can foster creativity, inclusivity, and resilience in the face of rapid change. Neodynamics redefines cultural systems and creative industries as adaptive, feedback-driven ecosystems capable of thriving in complexity. Through constructs like UFAP and SPARC, this framework ensures that cultural systems remain relevant, innovative, and inclusive, addressing the challenges of a rapidly evolving global landscape while fostering connection and shared meaning.

Homelessness and Poverty

Homelessness and poverty are complex, systemic issues shaped by economic inequality, structural discrimination, and resource misallocation. Traditional approaches to addressing these challenges often rely on static frameworks, such as predefined eligibility criteria or standardized assistance programs, which fail to adapt to the dynamic and multifaceted needs of individuals and communities. Neodynamics provides a transformative framework for embedding adaptability, feedback, and emergent coherence into strategies for reducing homelessness and poverty, ensuring that interventions are responsive, equitable, and sustainable.

1.   One-Size-Fits-All Solutions: Static assistance programs often apply rigid eligibility criteria and uniform interventions, neglecting the unique circumstances and needs of individuals experiencing poverty or homelessness.

2.   Inflexible Resource Allocation: Traditional systems allocate resources based on historical data or fixed priorities, failing to respond dynamically to shifts in demand or emerging crises.

3.   Fragmented Services: Disconnected agencies and organizations create gaps in service delivery, leaving individuals to navigate complex systems without adequate support.

These limitations highlight the need for adaptive, integrated systems capable of addressing the dynamic and systemic nature of homelessness and poverty. The Unified Field of Adaptive Potential (UFAP) models the spectrum of interventions and resource allocations available to reduce homelessness and poverty. By conceptualizing these issues as dynamic systems, UFAP enables policymakers, service providers, and community organizations to explore and implement adaptive strategies that evolve based on real-time feedback and emergent needs.

Applications include:

•   Dynamic Resource Allocation: UFAP allows governments and organizations to adjust resource distribution dynamically, ensuring that shelters, food banks, and financial aid programs meet evolving community demands.

•   Personalized Support Pathways: By mapping the unique needs and potential outcomes for individuals, UFAP enables the creation of tailored support plans that address housing, employment, healthcare, and social connections.

• Iterative Policy Design: Policymakers can use UFAP to model and refine interventions such as universal basic income (UBI), housing-first initiatives, or job training programs, ensuring alignment with local and systemic conditions.

The SPARC Framework operationalizes adaptability in poverty reduction strategies by embedding the Spectrum of Possibility and Recursive Choice into decision-making and service delivery processes. These mechanisms ensure that interventions are continuously refined based on feedback from individuals, communities, and systemic outcomes.

Key applications include:

1. Real-Time Service Adjustments: Recursive feedback enables shelters, aid programs, and outreach organizations to adjust their offerings dynamically, such as increasing mental health services during crises or reallocating housing resources based on weather conditions.

2. Decentralized Support Networks: The SPARC Framework fosters collaboration among community organizations,

government agencies, and nonprofits, enabling adaptive and decentralized responses to homelessness and poverty.

3. Iterative Impact Measurement: Feedback-driven processes allow stakeholders to evaluate the effectiveness of interventions continuously, refining strategies to improve outcomes and address emerging challenges.

Emergent coherence ensures that diverse actors—policymakers, service providers, activists, and community members—align dynamically to address homelessness and poverty. Neodynamics fosters this coherence by embedding feedback loops that integrate local actions with broader systemic strategies.

For example:

•   Community-Led Housing Solutions: A Neodynamic approach could involve decentralized networks of community groups and nonprofits that self-organize to create affordable housing, dynamically adjusting their efforts based on feedback from residents, resource availability, and changing needs.

This emergent alignment allows for holistic and sustainable solutions that address the root causes of homelessness and poverty while empowering communities. While applying Neodynamics to homelessness and poverty presents challenges, it also offers significant opportunities:

•   Scalability: Adaptive frameworks must be designed to scale across regions and populations, maintaining coherence while addressing local needs.

•   Equity: Interventions must prioritize equity, ensuring that marginalized groups receive support tailored to their unique circumstances and systemic barriers.

•   Data Integration: Effective adaptive systems require robust mechanisms for collecting, sharing, and analyzing data from diverse stakeholders and communities.

Despite these challenges, Neodynamics provides a pathway for transforming efforts to address homelessness and poverty into adaptive, equitable, and effective systems capable of fostering long-term change. Neodynamics redefines the fight against homelessness and poverty as an adaptive, feedback-driven process that integrates systemic insights with local action. By leveraging constructs like UFAP and SPARC, this framework ensures that interventions are responsive, inclusive, and sustainable, providing a roadmap for addressing these deeply entrenched social challenges with resilience and innovation.

Psychology and Mental Health

Psychology and mental health systems operate in dynamic and multifaceted contexts, shaped by individual variability, cultural influences, and societal pressures. Traditional psychological frameworks often rely on static diagnostic categories and predetermined treatment protocols, which fail to adapt to the evolving needs and circumstances of individuals. Neodynamics provides a framework for embedding adaptability, feedback-driven processes, and emergent coherence into mental health practices, enabling more personalized, responsive, and holistic care.

1.   Rigid Diagnostic Criteria: Traditional diagnostic frameworks, such as the DSM or ICD, categorize mental health conditions into static labels, neglecting the fluid and context-dependent nature of psychological experiences (Kupfer et al., 2013).

2.   Fixed Treatment Approaches: Therapies and interventions often follow predetermined protocols that may not evolve with a client's progress, setbacks, or changing circumstances.

3.   Fragmented Care Systems: Mental health services frequently operate in isolation, making it difficult to integrate feedback from multiple providers or account for the interplay of psychological, social, and physical health factors.

These limitations underscore the need for adaptive frameworks that reflect the complexity and dynamism of mental health. The Unified Field of Adaptive Potential (UFAP) provides a framework for mapping the adaptive pathways available to clients and practitioners. By conceptualizing mental health as a dynamic process, UFAP enables the exploration of diverse interventions and strategies that align with individual needs and systemic contexts. Applications include:

•   Dynamic Therapy Models: UFAP allows therapists to adapt their approaches in real time, integrating feedback from clients to refine techniques such as cognitive-behavioral therapy, mindfulness, or medication adjustments.

•   Iterative Diagnostic Processes: Instead of relying on fixed categories, UFAP supports a dynamic diagnostic framework that evolves as new insights about a client's experiences and conditions emerge.

• Personalized Recovery Pathways: By mapping the full range of supportive options, UFAP enables clients and practitioners to co-create adaptive recovery plans that encompass psychological, social, and physical health dimensions.

The SPARC Framework operationalizes adaptability in mental health systems by embedding the Spectrum of Possibility and Recursive Choice into diagnostic and therapeutic processes. These mechanisms ensure that care evolves continuously based on client feedback and systemic outcomes.

Key applications include:

1. Iterative Treatment Planning: Recursive feedback allows therapists and clients to adjust therapeutic goals and methods dynamically, reflecting progress, challenges, and changing life circumstances.

2. Collaborative Care Models: The SPARC Framework fosters decentralized collaboration among providers, integrating input from therapists, psychiatrists, primary care physicians, and community support networks.

3. Dynamic Crisis Intervention: Feedback-driven processes enable mental health services to respond adaptively to crises, such as

suicide prevention or emergency psychiatric care, by recalibrating interventions in real time.

Emergent coherence ensures that diverse actors and elements within mental health systems align dynamically to create cohesive and effective support networks. Neodynamics fosters this coherence by embedding feedback-driven processes that allow for seamless integration of individual care with broader systemic efforts.

For example:

• Community Mental Health Initiatives: A Neodynamic framework could enable decentralized networks of community organizations and healthcare providers to self-organize, dynamically addressing local mental health challenges while maintaining alignment with regional or national goals.

This emergent alignment enhances accessibility, equity, and sustainability in mental health systems.

Implementing Neodynamics in psychology and mental health presents challenges:

• Ethical Considerations: Adaptive frameworks must prioritize client consent, autonomy, and privacy, ensuring that iterative processes remain transparent and respectful.

• Cultural Sensitivity: Mental health interventions must be tailored to reflect diverse cultural values and practices, avoiding the imposition of one-size-fits-all solutions.

• Data Integration: Adaptive systems require robust mechanisms for collecting and analyzing feedback from clients, practitioners, and care networks.

Despite these challenges, Neodynamics offers transformative opportunities for mental health systems. By embedding adaptability and feedback into care models, it empowers clients and providers to navigate the complexity of mental health with resilience and innovation. Neodynamics reimagines mental health care as an adaptive, feedback-driven process that evolves continuously to meet the unique needs of individuals and communities. Through constructs like UFAP and SPARC, this framework provides a roadmap for creating responsive, holistic, and equitable mental health systems that thrive in complexity.

Race, Identity, and Social Justice

Race, identity, and social justice are deeply interconnected issues shaped by systemic inequities, cultural histories, and ongoing struggles for equity and liberation. Traditional approaches to addressing racial injustice often rely on static frameworks, such as predefined anti-discrimination policies or singular cultural narratives, which fail to account for the complexity and evolving nature of systemic oppression. Neodynamics offers a framework for embedding adaptability, feedback, and emergent coherence into efforts to dismantle racism and promote justice, ensuring that strategies are dynamic, inclusive, and sustainable

1.   Rigid Policies: Static anti-racism and social justice initiatives often implement top-down solutions that do not adapt to the specific needs of different communities or respond to emergent forms of systemic oppression.

2.   Performative Actions: Token gestures or symbolic commitments to diversity and inclusion often fail to address the underlying structures of inequality and exploitation.

3.   Fragmented Movements: Activist efforts frequently lack coherence and alignment, as diverse stakeholders pursue overlapping but uncoordinated strategies.

These limitations underscore the need for adaptive, systemic approaches that can evolve in response to the complexity of racial injustice. The Unified Field of Adaptive Potential (UFAP) provides a framework for exploring diverse pathways to racial equity and justice, enabling policymakers, educators, activists, and communities to identify and implement strategies that adapt dynamically to emergent challenges.

Applications include:

• Adaptive Equity Policies: UFAP supports the development of anti-racism policies that evolve based on real-time data, community input, and measurable outcomes, ensuring that interventions remain effective and equitable.

• Decentralized Activist Networks: UFAP enables grassroots movements to self-organize and adaptively address local manifestations of systemic racism while contributing to broader national and global efforts.

• Iterative Cultural Narratives: By modeling the adaptive landscape of cultural narratives, UFAP enables educators, artists, and media creators to refine anti-racism messages that resonate across diverse audiences and counter emergent resistance.

The SPARC Framework operationalizes adaptability in racial justice efforts, embedding the Spectrum of Possibility and Recursive Choice into activism, policymaking, and cultural change. These mechanisms ensure that strategies evolve continuously based on feedback and systemic impact.

Key applications include:

1.   Real-Time Community Responses: Recursive feedback allows communities to refine strategies dynamically, such as reallocating resources to support emerging needs like housing, healthcare, or legal advocacy.

2.   Dynamic Coalition Building: The SPARC Framework fosters collaboration among diverse movements, ensuring that localized actions align with broader anti-racism goals without erasing the unique needs of individual communities.

3.   Iterative Educational Models: Feedback-driven processes allow educators to adapt curricula and training programs to reflect the evolving understanding of systemic racism and intersectionality.

Emergent coherence ensures that the diverse actors and initiatives working toward racial equity align dynamically to create systemic change. Neodynamics fosters this coherence by embedding feedback-driven processes that integrate local efforts with national and global movements.

For example:

•   Adaptive Policy Ecosystems: A Neodynamic approach could enable municipalities, NGOs, and grassroots organizations to self-coordinate, dynamically adjusting their actions to address inequities in housing, policing, or education while aligning with long-term systemic reform goals.

This emergent alignment enhances the impact, scalability, and inclusivity of racial justice initiatives. Applying Neodynamics to racial justice presents challenges:

•   Cultural Resistance: Adaptive frameworks must navigate entrenched opposition and disinformation while maintaining integrity and focus.

•   Data and Representation: Ensuring that data collection and analysis reflect diverse experiences and voices is critical for avoiding bias and perpetuating inequities.

•   Sustained Engagement: Adaptive strategies must prioritize long-term momentum and capacity-building to ensure that movements remain resilient and impactful over time.

Despite these challenges, Neodynamics offers transformative opportunities for addressing racial injustice. By embedding adaptability, feedback, and emergent coherence into social justice efforts, it equips movements and institutions to dismantle systemic oppression and foster equity in complex and dynamic contexts. Neodynamics reimagines racial justice as an adaptive, feedback-driven process capable of evolving continuously to address the complexity of systemic inequities. Through constructs like UFAP and SPARC, this framework provides a powerful tool for creating dynamic, inclusive, and impactful strategies that align with the ongoing pursuit of equity and liberation.

Criminal Justice Reform

Criminal justice systems are often rigid, inequitable, and punitive, focusing on control and punishment rather than rehabilitation and systemic transformation. These systems disproportionately harm marginalized communities, perpetuate systemic biases, and fail to adapt to the dynamic and multifaceted nature of crime and social harm. Neodynamics offers a transformative framework for reimagining criminal justice as an adaptive, feedback-driven system that prioritizes equity, rehabilitation, and systemic coherence.

1.   Punitive Focus: Traditional criminal justice frameworks emphasize punishment over addressing root causes of crime, such as poverty, mental health, or systemic inequities.

2.   Systemic Bias: Static policies and practices often embed racial, socioeconomic, and gender biases, exacerbating inequality and undermining public trust.

3.   Inflexible Sentencing and Policies: Standardized sentencing guidelines and parole frameworks fail to adapt to individual circumstances or community needs, leading to injustices and inefficiencies.

These limitations highlight the need for adaptive systems that are responsive, equitable, and rehabilitative. The Unified Field of Adaptive Potential (UFAP) provides a framework for modeling dynamic, rehabilitative pathways within criminal justice systems. By conceptualizing justice as an adaptive process, UFAP enables policymakers, judges, and community organizations to explore and implement innovative solutions that evolve based on feedback and systemic outcomes.

Applications include:

• Dynamic Sentencing Models: UFAP enables courts to explore and adjust sentencing options dynamically, incorporating factors such as individual risk, community needs, and the potential for rehabilitation.

• Adaptive Rehabilitation Programs: UFAP supports the development of personalized rehabilitation plans that evolve based on participants' progress, addressing underlying causes of criminal behavior.

• Iterative Policy Development: Policymakers can use UFAP to model and refine interventions such as restorative justice programs, bail reform, or decarceration initiatives, ensuring that policies align with real-time data and long-term societal goals.

The SPARC Framework operationalizes adaptability in criminal justice systems by embedding the Spectrum of Possibility and Recursive Choice into decision-making, program design, and policy implementation. These mechanisms ensure that systems evolve continuously in response to feedback from individuals, communities, and systemic outcomes.

Key applications include:

1.   Real-Time Adjustments: Recursive feedback allows probation and parole systems to adjust conditions dynamically based on individuals' progress, ensuring support rather than punitive oversight.

2.   Collaborative Justice Models: The SPARC Framework fosters decentralized collaboration among courts, community organizations, and advocacy groups, enabling adaptive responses to systemic inequities and emerging challenges.

3.   Iterative Bias Mitigation: Feedback-driven processes enable criminal justice systems to identify and address systemic biases continuously, improving equity and trust.

Emergent coherence ensures that diverse stakeholders—judges, law enforcement, policymakers, advocates, and community members—align dynamically to create cohesive and effective systems of justice. Neodynamics fosters this coherence by embedding feedback loops that integrate local actions with broader reform goals.

For example:

•   Restorative Justice Ecosystems: A Neodynamic approach could involve decentralized networks of restorative justice programs that self-organize to address harm and repair relationships, dynamically adjusting their methods based on participant feedback and outcomes.

This emergent alignment creates a system that is equitable, rehabilitative, and responsive to the needs of individuals and communities.

Applying Neodynamics to criminal justice reform presents challenges:

• Cultural Resistance: Adaptive reforms may face resistance from traditional institutions or communities invested in punitive justice models.

• Data Integration and Privacy: Effective feedback mechanisms require robust data collection and analysis, balanced with the need to protect individual rights and privacy.

• Scalability: Ensuring that adaptive models can scale across jurisdictions while maintaining coherence and equity is critical for systemic transformation.

Despite these challenges, Neodynamics offers significant opportunities for reimagining criminal justice. By embedding adaptability, feedback, and emergent coherence into reform efforts, it provides a pathway for addressing systemic inequities, promoting rehabilitation, and fostering public trust. Neodynamics redefines criminal justice reform as an adaptive, feedback-driven process that prioritizes equity, rehabilitation, and systemic transformation. Through constructs like UFAP and SPARC, this framework ensures that justice systems evolve dynamically, addressing the complexities of crime and social harm while promoting fairness, trust, and societal well-being.

Anti-Capitalism and Systemic Transformation

Capitalism, as the dominant economic system of the modern era, has fueled technological innovation, global connectivity, and unprecedented economic growth. However, its inherent flaws—exploitation, inequality, environmental degradation, and systemic instability—have created crises that it is fundamentally ill-equipped to address. Neodynamics provides a framework for critically analyzing and reimagining systems beyond capitalism, embedding adaptability, feedback, and emergent coherence into structures that prioritize equity, sustainability, and collective well-being.

1.   Exploitation of Labor and Resources: Capitalism relies on the extraction of value from both human labor and natural

resources, often prioritizing profit over well-being. This results in widespread inequality and ecological destruction, as seen in exploitative labor practices and the commodification of essential resources like water and land (Marx, 1867).

2.   Inequity and Wealth Concentration: The capitalist system rewards capital ownership over labor, leading to an accumulation of wealth and power in the hands of a few. This concentration exacerbates inequality and undermines democratic governance, as elites exert disproportionate influence over policy and society (Piketty, 2013).

3.   Systemic Instability: The drive for perpetual growth creates cycles of boom and bust, destabilizing economies and societies. Financial crises, such as those of 1929 and 2008, are intrinsic to capitalism's speculative nature and its prioritization of short-term profit over long-term stability.

4.   Environmental Degradation: Capitalism's emphasis on growth and consumption directly conflicts with ecological limits. The system externalizes environmental costs, accelerating climate change, biodiversity loss, and resource depletion.

These flaws underscore the urgent need for alternative systems that prioritize collective well-being, ecological harmony, and adaptability. Neodynamics provides a critical lens through which to examine capitalism's limitations and a practical framework for designing adaptive systems that transcend its failures. By embedding adaptability, feedback, and emergent coherence into economic and social systems, Neodynamics reimagines how value is created, distributed, and sustained. The Unified Field of Adaptive Potential (UFAP) enables the exploration of diverse economic and social pathways beyond capitalism. By modeling the spectrum of potential actions, UFAP facilitates the design of systems that dynamically adapt to human and ecological needs.

Applications include:

• Commons-Based Resource Management: UFAP supports systems where resources are collectively managed through participatory governance, dynamically balancing ecological limits with community needs.

• Decentralized Value Creation: By exploring adaptive models of value creation, such as cooperative ownership or community currencies, UFAP enables systems that prioritize equity and collective well-being.

• Iterative Redistribution Frameworks: UFAP allows policymakers and communities to refine redistribution mechanisms dynamically, addressing systemic inequalities while fostering economic stability.

The SPARC Framework operationalizes adaptability in the transition from capitalism to more equitable systems. The Spectrum of Possibility fosters diverse approaches to resource allocation, governance, and production, while Recursive Choice ensures that these systems evolve based on real-time feedback.

Key applications include:

1. Iterative Redistribution Models: Recursive feedback enables adaptive systems of wealth and resource redistribution, such as universal basic income or participatory budgeting, that evolve in alignment with community needs and systemic conditions.

2. Decentralized Governance Structures: The SPARC Framework supports non-hierarchical governance models, enabling communities to self-organize and adaptively manage shared resources and decision-making processes.

3. Dynamic Degrowth Strategies: Feedback-driven processes allow societies to transition away from perpetual economic growth, prioritizing well-being and ecological stability while dynamically recalibrating policies to reflect changing conditions.

Emergent coherence ensures that decentralized, adaptive systems align dynamically with shared goals of equity, sustainability, and resilience. Neodynamics fosters this coherence by embedding feedback loops that integrate local actions with broader systemic objectives.

For example:

• Community-Led Economies: A Neodynamic framework enables decentralized networks of cooperatives and community organizations to self-organize, dynamically addressing local needs while aligning with regional or global sustainability goals.

This emergent alignment creates economic systems that are responsive, equitable, and ecologically sustainable, avoiding the pitfalls of centralized control or rigid ideologies.

While applying Neodynamics to anti-capitalist systems presents challenges, it also offers transformative opportunities:

- Resistance to Change: Transitioning from entrenched capitalist structures requires significant cultural and institutional shifts, which may face resistance from vested interests.

- Data and Coordination: Adaptive post-capitalist systems require robust mechanisms for collecting, analyzing, and sharing data across decentralized networks.

- Cultural Transformation: Overcoming the deep cultural imprint of capitalist values, such as individualism and consumerism, is essential for fostering systems based on cooperation and sustainability.

Despite these challenges, Neodynamics offers a powerful framework for transcending capitalism's limitations. By embedding adaptability, feedback, and coherence into economic systems, it provides a path toward equity, ecological harmony, and collective flourishing. Neodynamics critiques capitalism as a system fundamentally incapable of addressing the crises it perpetuates. By reimagining economic and social systems as adaptive, feedback-driven networks, Neodynamics offers a transformative vision for equitable and sustainable futures. Through constructs like UFAP and SPARC, it enables the design of systems that prioritize well-being, resilience, and ecological balance over profit, offering a powerful tool for systemic change.

Neodynamics is the Science of Our Shared Humanity

Neodynamics is more than a framework for adaptability and coherence—it is the science of humanity itself. By embracing the principles of potential and agency, Neodynamics reveals how we can navigate the complexity of our existence in ways that align with the natural rhythms of the universe. Its applications across diverse domains—from governance and healthcare to justice and sustainability—demonstrate that our systems, like ourselves, must be capable of learning, evolving, and flourishing in a world defined by interconnection and change.

At its heart, Neodynamics invites us to step away from the illusion of control that underpins so much of modern thought. Humanity's greatest failures—climate collapse, systemic inequities, and cultural alienation—stem not from a lack of ingenuity but from a refusal to live in tune with the systems that sustain us. By contrast, Neodynamics provides a blueprint for moving beyond extraction and domination toward a world of co-creation and balance.

Central to Neodynamics is the understanding that potential is not static—it is a dynamic, ever-evolving field of possibilities. This insight, embodied in the Unified Field of Adaptive Potential (UFAP), reshapes how we approach everything from individual growth to societal transformation. Rather than seeking rigid solutions or absolute answers, Neodynamics shows us how to explore, refine, and recalibrate in pursuit of our highest aspirations.

This focus on potential also underscores the deeply humanistic nature of Neodynamics. As individuals and as societies, we are defined not by fixed outcomes but by our capacity to adapt, learn, and create. By embedding adaptability into the systems we rely on, Neodynamics empowers us to unleash the fullest expression of human agency and ingenuity.

The principles of feedback, emergence, and coherence that underpin Neodynamics are not inventions—they are observations of how the universe itself operates. From the self-organizing patterns of ecosystems to the iterative learning processes of evolution, Neodynamics mirrors the dynamic, interconnected nature of life. By applying these principles to human systems, we are not imposing order; we are aligning with the natural processes that have sustained complexity and resilience for billions of years.

This alignment is not only practical but profoundly ethical. It challenges us to replace domination with partnership, extraction with regeneration, and control with collaboration. In doing so, Neodynamics offers a path toward a future where humanity flourishes not by conquering the world but by coexisting with it.

All the examples explored in this chapter—adaptive healthcare, equitable justice, sustainable economics, and more—point to a singular truth: Neodynamics is a science of empowerment. It enables systems that honor human dignity, foster creativity, and adapt to our evolving needs and dreams. By embracing adaptability, we unlock the potential to create systems that serve everyone, not just a privileged few; systems that balance individual freedom with collective well-being; systems that sustain life rather than deplete it.

In this sense, Neodynamics is not merely a tool for survival in a complex world—it is a roadmap for thriving. It empowers us to imagine and build a future where human flourishing is not constrained by outdated paradigms but unleashed by systems that reflect the richness and potential of life itself.

The journey through this chapter has demonstrated the universal applicability of Neodynamics. From dismantling systemic oppression to navigating global crises, it offers a unifying framework that transcends disciplines and domains. Its principles challenge us to rethink our assumptions, redesign our systems, and reclaim our agency as individuals and as a species.

As we conclude, one truth becomes clear: Neodynamics is not just a framework for change; it is a call to action. It asks us to reimagine humanity's role in the universe—not as masters of a fragile world but as participants in a vast, adaptive, and interdependent system. It shows us that our greatest potential lies not in controlling the world but in learning to live in harmony with it.

In embracing Neodynamics, we embrace the essence of what it means to be human: to adapt, to create, to connect, and to thrive. This is the science of humanity because it is the science of life itself. Let it guide us toward a future where potential is unlimited, agency is universal, and flourishing is shared by all.

Designing Adaptive Systems

The design of adaptive systems is the foundational step in operationalizing Neodynamics. Unlike traditional systems that prioritize stability under static conditions, adaptive systems are engineered to thrive in dynamic, uncertain environments. This section introduces a rigorous framework for designing systems that embody Neodynamic principles: adaptability, feedback-driven processes, and emergent coherence. These principles are implemented through structured methodologies, enabling systems to align with complex, evolving objectives.

Theoretical Foundations

Adaptive systems operate on the interplay between flexibility and stability, leveraging feedback and emergence to navigate uncertainty. The key theoretical foundations include:

1.   Feedback Loops: As articulated by Wiener (1948), feedback loops enable systems to self-regulate and adjust dynamically, ensuring alignment with environmental changes.

2.   Dynamic Stability: Stability is redefined not as rigidity but as the ability to maintain coherence while adapting to perturbations (Ashby, 1956).

3.   Emergent Coherence: Inspired by Prigogine's (1984) work, coherence arises from the interactions of decentralized components, creating order without centralized control.

Framework for Adaptive System Design

Designing adaptive systems involves five iterative steps: *defining objectives, mapping adaptive landscapes, embedding feedback loops, enabling decentralization,* and *iterative refinement.*

1.   Define System Objectives in a Dynamic Context

Adaptive systems must balance specificity with flexibility, ensuring that objectives can evolve alongside environmental changes.

•   Establish Core Functions: Identify the system's primary goals (e.g., resilience, efficiency, innovation).

•   Anticipate Evolution: Define objectives as dynamic targets, incorporating feedback mechanisms for continuous recalibration.

•   Example: In healthcare, an adaptive system prioritizes patient outcomes but evolves its protocols based on epidemiological trends and resource availability.

2.   Map the Adaptive Landscape

The adaptive landscape represents the range of possible states a system can occupy, shaped by constraints, opportunities, and feedback.

• Identify Constraints and Levers: Map factors influencing adaptability, such as resource limits or technological capabilities.

• Model Multi-Dimensional Pathways: Use computational tools to explore potential trajectories and trade-offs.

• Example: Climate adaptation systems model trade-offs between economic growth and ecological preservation, identifying pathways that optimize both.

3. Embed Feedback Loops

Feedback loops drive system adaptation, enabling real-time responses to changes and disruptions.

• Real-Time Monitoring: Sensors, analytics, or human oversight capture critical data inputs.

• Recursive Adjustment: Feedback informs iterative recalibration of strategies, ensuring alignment with dynamic goals.

• Example: Autonomous vehicles use real-time feedback from sensors to navigate changing traffic conditions and optimize routes dynamically.

4.   Enable Decentralized Decision-Making

Decentralization enhances resilience and responsiveness by distributing decision-making authority across system components.

- Empower Local Autonomy: Allow subsystems to act independently within defined parameters.

- Coordinate Through Emergent Coherence: Align decentralized actions with overarching goals via shared incentives and feedback loops.

- Example: Decentralized energy grids optimize power distribution locally while ensuring system-wide efficiency.

5.   Iterative Refinement

Adaptive systems are never static; their design must facilitate continuous evolution through iterative testing and learning.

- Modular Architecture: Systems are built in modular components that can evolve independently.

- Stakeholder Feedback: Engage users and stakeholders in refining system behavior and objectives.

- Example: AI systems use iterative training cycles, incorporating user interactions to improve performance over time.

Practical Applications

1.   Urban Planning

Adaptive city systems integrate feedback from traffic patterns, resource use, and citizen inputs to optimize infrastructure dynamically.

- Example: Smart cities use IoT sensors to adjust public transport schedules in response to real-time demand.

2.   Healthcare Systems

Personalized medicine systems dynamically adapt to patient responses and health trends, ensuring optimal care.

- Example: AI-driven diagnostics refine treatment recommendations based on iterative data from patient outcomes.

3.   Supply Chains

Adaptive logistics systems respond to disruptions by recalibrating inventory flows and delivery schedules.

- Example: During crises, adaptive supply chains reallocate resources dynamically to meet changing demand.

Designing adaptive systems requires a paradigm shift from static, linear approaches to dynamic, iterative methodologies. By embedding Neodynamic principles—feedback integration, emergent coherence, and adaptability—into system design, organizations can create frameworks capable of thriving in complex and uncertain environments. This foundation sets the stage for testing and refining adaptive systems.

Simulation and Modeling

Simulation and modeling are essential tools for operationalizing Neodynamics, offering controlled environments to test theoretical constructs, refine adaptive behaviors, and identify emergent properties. By replicating complex systems, simulations reveal insights into feedback dynamics, coherence formation, and systemic adaptability. This section expands the scope of simulation methodologies, provides domain-specific examples, and emphasizes the role of feedback loops in refining Neodynamic systems.

1. Expanding Simulation Methodologies

Agent-Based Models (ABM)

- Overview: ABM simulates interactions between individual agents (e.g., people, machines, organizations) operating under simple rules. The emergent behavior of these agents offers insights into systemic dynamics.

- Strengths:

o Captures localized decision-making and its influence on macro-level phenomena.

o Models decentralized systems where emergent coherence is critical.

- Example:

o Urban Traffic Systems: Agents represent individual vehicles adapting routes based on real-time congestion. Emergent patterns reveal self-organization or chaotic bottlenecks, guiding infrastructure planning.

System Dynamics (SD)

- Overview: SD models systems as interconnected stocks, flows, and feedback loops. It is particularly suited for long-term trend analysis and high-level system behaviors.

- Strengths:

o Highlights delays, bottlenecks, and reinforcing feedback effects.

o Simulates the impact of policy decisions on systemic outcomes.

- Example:

- o Climate Adaptation: SD evaluates how carbon pricing policies influence emissions, energy use, and economic growth over decades, helping policymakers balance competing objectives.

Hybrid Simulations

- Overview: Combines ABM and SD to bridge micro-level interactions with macro-level trends.

- Strengths:

  - o Captures the nuanced interplay between localized decisions and systemic outcomes.

  - o Addresses multi-scale dynamics, essential for complex systems like global supply chains.

- Example:

  - o Healthcare Systems: ABM models individual patient behaviors (e.g., vaccination uptake), while SD tracks resource allocation across hospitals. Insights help optimize strategies during pandemics.

2. Domain-Specific Applications

Disaster Response Systems

- Scenario: Simulating emergency responses during hurricanes or wildfires.

- Simulation Details:

  o ABM: Agents represent responders, households, and infrastructure nodes, adapting routes and priorities based on real-time conditions.

  o SD: Models long-term resource depletion, recovery rates, and systemic resilience.

  o Hybrid: Combines localized evacuation modeling with regional resource allocation.

- Insights:

  o Identifies critical infrastructure vulnerabilities, such as single points of failure in evacuation routes.

  o Suggests staging strategies for emergency supplies and personnel to maximize impact.

Adaptive Agriculture

- Scenario: Modeling farming systems that respond dynamically to climate variability and market demand.

- Simulation Details:

- o ABM: Simulates farmer decision-making based on weather predictions, soil quality, and market prices.

- o SD: Tracks regional trends in water use, crop yields, and soil health over multiple growing seasons.

- o Hybrid: Integrates individual farm-level adaptations with broader regional food security dynamics.

- Insights:

  - o Demonstrates trade-offs between maximizing short-term profit and ensuring long-term sustainability.

  - o Recommends crop diversification and resource-sharing networks to enhance resilience.

Public Health Systems

- Scenario: Simulating disease outbreaks and adaptive healthcare responses.

- Simulation Details:

  - o ABM: Models individual behaviors, such as vaccination, social distancing, and quarantine compliance.

  - o SD: Tracks healthcare capacity, resource allocation, and immunity trends over time.

- o Hybrid: Combines individual behaviors with systemic healthcare responses to evaluate adaptive policies.

- Insights:

  - o Identifies tipping points for herd immunity and resource overload.

  - o Suggests optimal vaccination strategies and resource deployment to minimize fatalities.

3. Feedback Loops in Simulations

Integrating feedback mechanisms into simulations is essential for reflecting the dynamic, real-time adaptability central to Neodynamics. Feedback loops ensure that simulations capture both the immediate and long-term consequences of system actions.

Dynamic Feedback Integration

1. Real-Time Data Updates:

   - o Simulations adjust parameters dynamically as new data becomes available.

   - o Example: In adaptive traffic models, feedback from congestion sensors reroutes vehicles in real time, preventing bottlenecks and optimizing flow.

2. Feedback-Driven Adaptations:

- o Systems continuously learn from feedback, refining their responses iteratively.

- o Example: Healthcare simulations integrate patient recovery data to adjust treatment protocols dynamically during pandemics.

Layered Feedback Loops

1. Micro-Level Feedback:

   - o Captures localized system responses to changes in the immediate environment.

   - o Example: Farmers in adaptive agriculture models adjust irrigation based on real-time soil moisture levels.

2. Macro-Level Feedback:

   - o Evaluates systemic impacts and long-term trends resulting from aggregated micro-level decisions.

   - o Example: Climate adaptation simulations track how regional energy policies influence global emission trajectories.

Challenges and Solutions in Feedback Modeling

1. Challenge: Delayed Feedback Effects

o   Delayed responses in systems (e.g., policy changes) may destabilize simulations.

o   Solution: Incorporate historical data to predict delayed effects and preemptively adjust feedback mechanisms.

2. Challenge: Feedback Oscillations

o   Overcorrecting based on amplified feedback can lead to instability.

o   Solution: Introduce damping factors that moderate responses to extreme feedback signals.

4. Iterative Refinement Through Simulation

Simulations are not static experiments; they evolve iteratively, improving accuracy and applicability with each cycle.

Phase 1: Initial Model Development

- Build a foundational model reflecting the key variables and interactions of the system.

- Example: A baseline energy grid simulation that models renewable energy inputs and demand fluctuations.

Phase 2: Scenario Testing

- Stress-test the system across various scenarios to evaluate its resilience and adaptability.

- Example: Simulate drought conditions in adaptive agriculture models to identify vulnerabilities in water usage.

Phase 3: Stakeholder Feedback Integration

- Incorporate qualitative and quantitative feedback from stakeholders to refine the model.

- Example: Urban planning simulations integrate resident feedback on traffic flow and green space allocation.

Phase 4: Adaptive Iteration

- Use feedback from initial simulations to iteratively refine parameters and rules.

- Example: Adaptive healthcare simulations adjust triage protocols based on outcomes from previous iterations.

5. Role of Machine Learning in Neodynamic Simulations

Reinforcement Learning

- Systems learn optimal strategies through trial-and-error interactions with the simulation environment.

- Example: Adaptive AI systems in logistics simulations refine delivery routes based on feedback from previous disruptions.

Predictive Analytics

- Machine learning predicts future states of the system, enhancing its ability to preemptively adapt.

- Example: Climate models use predictive analytics to forecast extreme weather patterns and recommend adaptive infrastructure investments.

Multi-Agent Learning

- Agents within ABM simulations learn collaboratively or competitively, mimicking real-world behaviors.

- Example: Autonomous vehicles in traffic simulations learn to optimize routes while avoiding collisions.

Simulation and modeling are indispensable for validating and refining Neodynamic systems. By leveraging advanced methodologies, integrating feedback loops, and iteratively refining models, simulations bring Neodynamics closer to real-world applicability. These tools reveal the interplay between local actions and systemic outcomes, ensuring that adaptive systems can thrive in complexity.

Metrics for Evaluation

Metrics for evaluating adaptive systems are essential for assessing their alignment with Neodynamic principles—adaptability, feedback integration, and emergent coherence. Traditional metrics often fall short in capturing the dynamic, non-linear, and context-dependent nature of adaptive systems. This section introduces a comprehensive framework for measuring the success of Neodynamic systems, emphasizing both quantitative and qualitative indicators.

Core Evaluation Criteria

1.   Adaptability

•   Definition: The system's ability to adjust its behavior in response to environmental changes without compromising its core objectives.

•   Indicators:

•   Response Time: How quickly the system adjusts to disruptions.

•   Range of Responses: The diversity and relevance of adaptive strategies deployed.

•   Resilience Index: The system's ability to recover and maintain functionality after a disturbance.

- Example: In healthcare, an adaptive system adjusts its resource allocation in real time to address patient surges during pandemics.

2. Feedback Integration

- Definition: The system's capacity to incorporate internal and external feedback effectively into its operations and decision-making processes.

- Indicators:

- Signal Accuracy: The ability to differentiate between relevant and irrelevant feedback.

- Feedback Utilization: The degree to which feedback informs system adjustments.

- System Stability: Whether feedback loops stabilize the system or cause oscillations.

- Example: In supply chains, feedback from inventory levels triggers dynamic reordering, optimizing stock without overcorrection.

3. Emergent Coherence

- Definition: The alignment of decentralized system components to create functional, systemic order without centralized control.

- Indicators:

- Goal Alignment: The degree to which local decisions support global objectives.

- Pattern Formation: Evidence of emergent structures or behaviors that enhance system efficiency.

- Resilience through Redundancy: The system's ability to maintain coherence under stress through distributed redundancies.

- Example: In decentralized energy grids, coherence is measured by the seamless synchronization of local power nodes with the overarching grid.

Specific Metrics for Neodynamic Systems

1. Dynamic Stability Index (DSI)

- Purpose: Quantifies the system's ability to balance stability and adaptability.

- Components:

- Speed of response to changes.

- Degree of functional continuity during disruptions.

- Applications: Evaluate financial markets for their ability to recover from economic shocks without destabilizing.

2. Feedback Efficacy Ratio (FER)

- Purpose: Measures how effectively feedback loops improve system performance.

- Components:

- Ratio of feedback-driven changes to total system inputs.

- Stability of adjustments over time.

- Applications: Assess feedback integration in AI systems, ensuring that user inputs lead to meaningful system improvements.

3. Emergent Coherence Score (ECS)

- Purpose: Evaluates the alignment of decentralized components within the system.

- Components:

- Degree of alignment with overarching goals.

- Consistency of patterns across decentralized nodes.

- Applications: Assess coherence in adaptive governance systems, where local policies must align with national strategies.

4. Adaptation Energy Metric (AEM)

- Purpose: Tracks the resource expenditure required for system adaptation.

- Components:

- Efficiency of resource allocation.

- Balance between energy expenditure and adaptive gains.

- Applications: Monitor ecological systems for resource-efficient adaptation to environmental changes.

Qualitative Metrics

Quantitative indicators are complemented by qualitative metrics to capture the nuanced, context-dependent nature of adaptive systems:

- Stakeholder Feedback: Surveys and interviews with users, operators, and stakeholders to assess system performance and perceived equity.

- Case Narrative Analysis: Detailed reviews of specific adaptation events to identify emergent patterns and lessons learned.

- Context-Specific Outcomes: Tailored metrics reflecting unique objectives, such as community well-being or ecological restoration.

Challenges in Measuring Adaptive Systems

1. Dynamic Baselines

- Issue: Traditional baselines may not apply to systems that continuously evolve.

- Solution: Use rolling baselines that adjust dynamically to reflect system changes over time.

2.   Interdependencies

- Issue: Metrics must account for the interactions between subsystems, which can complicate attribution of outcomes.

- Solution: Employ network analysis to measure system-wide impacts of localized changes.

3.   Ethical Considerations

- Issue: Metrics must balance efficiency with equity, avoiding trade-offs that disproportionately harm marginalized groups.

- Solution: Integrate equity-weighted metrics to ensure fair distribution of system benefits.

Practical Applications of Metrics

1.   Generative AI

- Metrics such as FER and ECS ensure that AI models adapt dynamically to user inputs while maintaining coherence in outputs.

2.   Urban Systems

•   DSI and AEM measure how cities balance adaptability to changing demands with efficient resource allocation.

3.   Global Governance

•   ECS and stakeholder feedback evaluate the alignment of decentralized international efforts in addressing global crises.

Metrics for evaluating adaptive systems provide a structured way to assess their alignment with Neodynamic principles. By combining quantitative and qualitative approaches, these metrics capture the dynamic, context-sensitive nature of adaptability, feedback integration, and emergent coherence. This foundation ensures that Neodynamic systems are rigorously evaluated, driving continuous improvement and accountability. Next, we explore Interdisciplinary Collaboration, which is essential for implementing these systems effectively across domains.

Interdisciplinary Collaboration

The complexity of modern challenges requires solutions that transcend disciplinary boundaries. Neodynamics, with its emphasis on adaptability, feedback, and emergent coherence, inherently relies on insights from multiple fields. Interdisciplinary collaboration is not only desirable but essential for implementing adaptive systems effectively. This section explores strategies for fostering interdisciplinary integration, addressing challenges, and illustrating the transformative potential of cross-domain collaboration.

The Imperative for Interdisciplinary Collaboration

1. Complex Problems Demand Holistic Solutions

• Argument: Many real-world challenges—such as climate change, healthcare disparities, and global governance—are inherently multifaceted. No single discipline can fully address their intricacies (Rockström et al., 2009).

• Example: Climate resilience requires expertise in ecology, engineering, economics, and political science to design sustainable solutions that balance environmental preservation with human development.

2. Innovation Through Perspective Diversity

• Argument: Collaborative teams composed of diverse disciplinary backgrounds are more likely to generate innovative solutions, as differing perspectives highlight blind spots and uncover novel pathways (Rhoten & Parker, 2004).

- Example: In adaptive AI design, collaboration between computer scientists, ethicists, and behavioral psychologists ensures that systems are technically robust, ethically sound, and user-aligned.

3.   Systemic Coherence

- Argument: Interdisciplinary collaboration aligns local and global objectives, fostering systemic coherence across scales.

- Example: Decentralized governance models integrate local cultural knowledge with global best practices, ensuring policies are both context-sensitive and globally coordinated.

Frameworks for Collaboration

1.   Integrated Design Teams

- Structure: Teams bring together experts from relevant disciplines, ensuring diverse perspectives and expertise.

- Key Practices:

- Role Specialization: Each member contributes unique expertise, such as data analysis, policy design, or stakeholder engagement.

- Cross-Training: Team members are cross-trained in complementary disciplines to enhance communication and integration.

- Example: In urban planning, integrated teams include architects, urban ecologists, sociologists, and transportation engineers, ensuring adaptive city systems address ecological and social needs simultaneously.

2.   Collaborative Platforms

- Purpose: Digital tools facilitate communication, data sharing, and joint problem-solving across disciplinary boundaries.

- Examples:

- Shared Modeling Environments: Tools like AnyLogic enable teams to co-create hybrid simulations, integrating agent-based models with system dynamics.

- Collaborative Dashboards: Real-time visualization platforms aggregate inputs from multiple disciplines, aiding joint decision-making.

- Applications: In adaptive supply chain management, platforms allow logisticians, economists, and technologists to model disruptions and optimize solutions collaboratively.

3.   Boundary-Spanning Roles

•   Definition: Individuals trained to navigate and integrate insights across disciplines serve as translators and mediators in collaborative settings.

•   Key Practices:

•   Systems Thinking Training: Professionals are trained in foundational systems principles to bridge disciplinary divides.

•   Institutional Support: Organizations establish roles specifically for interdisciplinary integration, such as Chief Systems Officer or Collaboration Architect.

•   Example: In healthcare systems, boundary spanners coordinate between clinical staff, data scientists, and administrators to implement adaptive patient care models.

Challenges in Collaboration

1.   Communication Barriers

•   Issue: Differences in terminology, methodologies, and priorities across disciplines can create misunderstandings and inefficiencies.

•   Solution: Develop a shared language through workshops, glossaries, and iterative dialogue.

- Example: In climate adaptation projects, ecologists and economists align by agreeing on standardized metrics for ecosystem services valuation.

2. Conflicting Objectives

- Issue: Disciplinary goals may diverge, leading to tension and misaligned efforts.

- Solution: Employ iterative feedback loops to identify and resolve conflicts early, ensuring alignment with overarching system objectives.

- Example: In adaptive governance, conflict between short-term economic goals and long-term sustainability can be resolved through participatory decision-making frameworks.

3. Resource Allocation

- Issue: Interdisciplinary projects often face funding and logistical constraints, particularly when spanning institutional boundaries.

- Solution: Advocate for cross-sectoral funding mechanisms, such as public-private partnerships or interdisciplinary research grants.

- Example: The European Union's Horizon 2020 program funds interdisciplinary research addressing global challenges, such as renewable energy systems.

Applications of Interdisciplinary Collaboration

1. Healthcare

- Collaboration: Medical professionals, behavioral scientists, and data engineers design adaptive healthcare systems that integrate real-time data with patient-centered care.

- Outcome: Reduced misdiagnoses and improved patient outcomes through dynamic resource allocation and personalized treatment plans.

2. Climate Resilience

- Collaboration: Ecologists, civil engineers, and policymakers co-create adaptive infrastructure that balances ecological preservation with human needs.

- Outcome: Flood-resistant cities with green infrastructure that adapts dynamically to changing climate conditions.

3. Generative AI

- Collaboration: Computer scientists, ethicists, and artists develop adaptive AI systems that generate creative content while respecting cultural and ethical norms.

- Outcome: AI platforms that adapt dynamically to user feedback while maintaining coherence and originality.

Interdisciplinary collaboration is a cornerstone of Neodynamics, enabling the integration of diverse knowledge and expertise into adaptive system design. By fostering shared goals, employing collaborative platforms, and addressing challenges proactively, interdisciplinary teams can unlock the full potential of Neodynamic systems. This foundation ensures that solutions are holistic, innovative, and aligned with the complexity of real-world challenges.

Conclusion

The methodologies explored here are the bridge between the theoretical foundations of Neodynamics and its tangible impact on the world. They are the tools that bring adaptability, coherence, and feedback to life, shaping systems that can navigate complexity with purpose and precision. From the elegance of simulations to the rigor of evaluation metrics, these methods are not just technical—they are the embodiment of Neodynamics' philosophy, turning potential into reality.

What makes this work vital is its recursive nature. These methodologies are not static blueprints but evolving frameworks, designed to learn, adapt, and refine themselves. They mirror the systems they create, growing stronger through iteration and feedback. This is the essence of Neodynamics—not control, but alignment; not rigidity, but resilience.

The ideas and strategies here are the foundation for what comes next, the starting point for designing systems that thrive, empower, and sustain. They remind us that the science of adaptation is also the art of potential, where theory and practice converge to shape a future defined by creativity, balance, and flourishing.

Challenges and Criticisms

No framework, including Neodynamics, is without limitations. Its ambitious goals—to unify adaptability, feedback, and emergent coherence—invite theoretical and practical challenges. This chapter critically examines the critiques of Neodynamics, from potential theoretical weaknesses to implementation barriers and ethical concerns. By addressing these issues head-on, Neodynamics not only refines its framework but also demonstrates its commitment to rigorous self-evaluation and continuous improvement.

Theoretical Limitations

Criticism 1: Ambiguity in Emergent Coherence

Emergent coherence, a core principle of Neodynamics, is criticized for being conceptually abstract and difficult to operationalize. Critics question whether it can be reliably measured or achieved in complex systems.

Emergence relies on the spontaneous alignment of decentralized components, which may not always converge predictably. Measurement challenges arise in determining when coherence has been achieved and how it contributes to systemic goals. Neodynamics formalizes emergent coherence through metrics like the Coherence Alignment Metric (CAM), which quantifies alignment and pattern formation in decentralized systems. Practical applications, such as decentralized energy grids, illustrate that emergent coherence is achievable when subsystems operate under shared incentives and feedback loops.

Further Refinement: Research into mathematical modeling of emergence, such as graph theory and network science, offers pathways to quantify coherence more rigorously (Barabási, 2016).

Criticism 2: Overreliance on Feedback Loops

Feedback loops are essential to Neodynamics, yet critics argue they may create unintended consequences, such as oscillations, overcorrections, or delays in adaptation. Delayed feedback integration may cause systems to respond to outdated conditions, leading to maladaptive behaviors. Positive feedback, if unchecked, risks destabilizing the system (e.g., financial bubbles).

Neodynamics incorporates multi-layered feedback systems, combining real-time local feedback with slower, systemic feedback to balance stability and adaptability. Negative feedback loops are explicitly embedded to counteract runaway positive feedback, ensuring stability under dynamic conditions. Autonomous vehicles use layered feedback: immediate data from sensors for navigation and aggregated traffic data for broader route optimization. This prevents oscillatory responses to transient disruptions.

Criticism 3: Lack of Predictive Precision

Neodynamics emphasizes probabilistic and adaptive approaches over deterministic predictions. Critics argue that this sacrifices clarity and certainty, particularly in high-stakes systems. Probabilistic models may lack the predictive precision required for time-sensitive or high-risk environments, such as disaster management. Adaptation over prediction might result in slower responses to critical events.

Neodynamics supplements probabilistic modeling with scenario-based planning, enabling systems to anticipate a range of possible outcomes. Hybrid frameworks combine deterministic elements for short-term predictability with adaptive mechanisms for long-term resilience. Adaptive climate models use deterministic inputs for immediate weather forecasts while relying on probabilistic models for long-term policy planning.

Criticism 4: Epistemological Challenges

The dynamic and iterative nature of Neodynamics raises questions about the epistemological grounding of its principles. Critics may argue that it lacks a stable theoretical foundation. Constant adaptation and recalibration challenge traditional notions of knowledge as stable or definitive. Systems that rely on continual learning may fail in environments where feedback is ambiguous or insufficient.

Neodynamics adopts a Bayesian epistemology, emphasizing learning through evidence accumulation and iterative hypothesis testing. By incorporating redundancy and exploratory mechanisms, Neodynamics ensures that systems can operate effectively even under incomplete or uncertain feedback. This epistemological flexibility aligns with contemporary science, which increasingly views knowledge as provisional and evolving (Popper, 1959).

Theoretical critiques of Neodynamics, while valid, serve to refine and strengthen its principles. By addressing concerns about emergent coherence, feedback dynamics, predictive precision, and epistemology, Neodynamics evolves into a more robust and defensible framework. These refinements not only enhance its theoretical rigor but also solidify its practical applicability across domains.

Implementation Barriers

Implementing Neodynamics in real-world systems faces significant barriers, stemming from technological limitations, institutional resistance, and broader systemic challenges. These barriers highlight the complexity of transitioning from theory to practice and demand innovative approaches to overcome. This section critically examines these obstacles and offers practical strategies to address them.

Barrier 1: Technological Constraints

Adaptive systems require advanced computational tools, real-time data integration, and scalable architectures. These technological demands pose a challenge, particularly in resource-constrained environments or emerging economies. Adaptive systems, particularly those involving hybrid simulations or multi-layered feedback loops, demand significant computational resources. Small organizations or governments with limited access to high-performance computing (HPC) may struggle to implement Neodynamic systems effectively.

Systems are developed in scalable modules, allowing organizations to implement basic functionalities before expanding. Leveraging distributed cloud services reduces the barrier to accessing computational power. Decentralized healthcare systems in rural regions can start with telemedicine platforms before integrating AI-driven diagnostics.

The efficacy of Neodynamic systems hinges on reliable, high-quality data, which may be unavailable or fragmented in some contexts. Poor data can lead to erroneous feedback loops, maladaptive responses, or systemic failures.

Employing redundant data sources and cross-validation techniques ensures robustness against incomplete or erroneous datasets. Machine learning models can generate synthetic datasets to fill gaps, particularly in training phases. During the COVID-19 pandemic, missing data on case numbers was mitigated by predictive models that filled gaps using historical trends and demographic correlations.

Adaptive systems, by nature of their interconnectivity, are vulnerable to cyberattacks and data breaches. Compromised feedback loops can destabilize systems or expose sensitive information.

Systems embed self-adjusting security measures, such as anomaly detection algorithms powered by AI. Distributed designs minimize single points of failure, enhancing resilience against attacks. Blockchain technology in adaptive supply chains ensures secure, transparent transactions even during disruptions.

Barrier 2: Institutional Resistance

Institutional inertia and entrenched power structures often resist the decentralization and iterative decision-making central to Neodynamics. Decentralized systems challenge traditional hierarchical models, particularly in industries or governments accustomed to centralized authority. Without buy-in from leadership, implementation efforts may stall or face active opposition.

Pilot projects illustrating the benefits of decentralization build confidence and reduce resistance. Hybrid governance structures balance decentralized decision-making with centralized oversight during the transition phase. In adaptive governance models, local councils handle context-specific issues while central authorities monitor overarching coherence.

Stakeholders may resist adaptive changes that appear to threaten their vested interests or create short-term risks. Misaligned incentives hinder collaboration and adoption. Neodynamics employs incentive structures that align individual gains with collective benefits, minimizing conflict. Real-time dashboards display system performance, building trust through transparency. In adaptive energy grids, profit-sharing models encourage local producers to align with global efficiency goals.

Organizations often operate in silos, with limited collaboration across departments or sectors. This fragmentation undermines the integration required for adaptive systems. Dedicated roles, such as systems integrators, bridge silos by facilitating cross-departmental collaboration. Digital platforms aggregate and disseminate information across silos, fostering unified action. In disaster response, shared platforms integrate real-time data from health, logistics, and emergency services to enable cohesive action.

Barrier 3: Systemic Challenges

Neodynamic systems do not operate in isolation—they are embedded within larger societal, economic, and ecological contexts. These systemic challenges present some of the most significant obstacles to successful implementation. Societal norms and values may conflict with the principles of adaptability and feedback-driven change. Communities often resist systemic changes that challenge traditional structures or require significant behavioral shifts. Without cultural alignment, even technically sound systems may face rejection or underutilization.

Engaging stakeholders in the design process ensures cultural alignment and fosters a sense of ownership. Gradual adoption strategies minimize disruption, allowing communities to adapt over time. In urban planning, adaptive systems for public transport succeed when local preferences and patterns are integrated into the design.

Adaptive systems may inadvertently deepen economic disparities if resource allocation favors affluent regions or groups with better access to technology. Inequitable access undermines the universal applicability of Neodynamics, perpetuating systemic injustices.

Neodynamic systems embed equity-based metrics that prioritize underserved communities. Key technologies are made open-source, reducing barriers to entry for resource-constrained actors. Decentralized energy grids redistribute surplus power from wealthier nodes to underserved areas, ensuring equitable access.

Adaptive systems may place additional demands on finite resources, particularly in ecologically fragile contexts. High-tech implementations, such as AI-driven analytics, can also have a significant carbon footprint. Unsustainable designs risk ecological degradation, contradicting the principles of long-term adaptability.

Neodynamic systems integrate principles of circularity, minimizing waste and actively restoring ecosystems. Systems are evaluated based on their ecological impact, ensuring alignment with environmental goals. Adaptive agriculture models prioritize water-efficient crops and regenerative soil practices to balance food security with ecological preservation.

Turning Barriers into Opportunities

Challenges in implementation are not just obstacles; they are opportunities to refine and strengthen Neodynamic principles. Addressing these barriers forces systems to evolve, becoming more resilient and aligned with real-world complexities. Partnerships between governments, private sectors, and communities foster creative solutions to systemic challenges. Regulatory frameworks explicitly support the adoption of adaptive technologies, ensuring institutional alignment. Platforms for sharing lessons and best practices accelerate progress and prevent repeated mistakes.

Implementation barriers highlight the complexity and ambition of Neodynamics. By addressing technological, institutional, and systemic challenges with pragmatic solutions, adaptive systems can move from conceptual frameworks to impactful real-world applications. These barriers are not a limitation of Neodynamics but a testament to its potential for transformative change.

Ethical Considerations

Neodynamics, as a framework for designing adaptive systems, inherently carries profound ethical responsibilities. By embedding feedback loops and fostering emergent coherence, it influences decision-making processes and outcomes that directly impact individuals, communities, and ecosystems. This section expands on ethical considerations, addressing challenges like algorithmic bias, privacy concerns, equity, and the unintended consequences of adaptation. Specific case studies and practical solutions ground these challenges in real-world contexts.

1. Algorithmic Bias and Equity

Adaptive systems rely heavily on data and algorithms, which can unintentionally perpetuate or exacerbate existing biases.

1. Bias in Training Data:

   o Historical inequities encoded in training datasets influence decision-making processes.

   o Example: In predictive policing, biased datasets lead to disproportionate surveillance and enforcement in marginalized communities.

2. Feedback Loops Amplifying Inequities:

   o Adaptive systems that learn from biased feedback risk reinforcing discriminatory outcomes.

   o Example: Hiring platforms relying on historical recruitment data may inadvertently deprioritize underrepresented candidates.

Impact

- Undermines public trust in adaptive systems.
- Marginalizes vulnerable populations, perpetuating systemic inequities.

Solutions

1. Bias Auditing and Correction:

   o Regularly audit datasets and algorithms to identify and mitigate biases.

   o Example: Healthcare systems use demographic-weighted datasets to ensure fairness in diagnostic AI tools.

2. Equity-Driven Feedback Mechanisms:

   o Embed equity metrics into feedback loops, dynamically recalibrating outcomes to correct systemic imbalances.

   o Example: Education platforms prioritize underserved schools in adaptive resource allocation.

3. Stakeholder Inclusion:

   o Engage diverse stakeholders in system design to ensure broad representation and inclusivity.

   o Example: Participatory governance in adaptive urban planning includes input from underrepresented communities.

2. Privacy and Surveillance

Real-time data collection, a cornerstone of adaptive systems, raises significant privacy concerns and risks of surveillance abuse.

1. Data Exploitation:

   o Continuous monitoring risks misuse of personal data, particularly in healthcare and governance contexts.

   o Example: Contact-tracing apps for pandemics raise concerns about long-term data retention and unauthorized sharing.

2. Surveillance Infrastructure:

   o Adaptive systems can unintentionally create pervasive surveillance networks.

   o Example: Smart cities using sensor networks for optimization may inadvertently compromise individual privacy.

Impact

- Violates fundamental rights to privacy and autonomy.

- Erodes public trust, hindering system adoption.

Solutions

1. Privacy-by-Design Principles:

- o Design systems with built-in safeguards, such as anonymization and data minimization.

- o Example: An adaptive energy grid anonymizes household data while optimizing consumption patterns.

2. Decentralized Data Management:

- o Employ decentralized architectures (e.g., blockchain) to minimize risks of centralized data breaches.

- o Example: In healthcare, decentralized patient records enhance privacy while enabling adaptive care.

3. Transparent Governance:

- o Establish clear data policies that inform users about how their data is collected, stored, and used.

- o Example: Urban systems disclose data usage in participatory dashboards, fostering transparency and trust.

3. Overadaptation and Context-Specific Risks

Systems overly optimized for specific contexts or short-term goals may lose flexibility or create unintended consequences.

1. Short-Term Overfitting:

- o Systems adapt too narrowly to current conditions, neglecting long-term sustainability.

- o Example: Adaptive agriculture systems prioritize immediate yield gains, exacerbating soil degradation.

2. Unanticipated Trade-Offs:

- o Feedback loops may optimize one metric at the expense of others, causing systemic imbalances.

- o Example: Urban transport systems designed to minimize travel times may increase pollution in residential neighborhoods.

Impact

- Creates fragility in dynamic environments.

- Undermines system resilience and long-term adaptability.

Solutions

1. Exploration-Exploitation Balance:

- o Ensure systems balance optimizing known solutions with exploring novel strategies.

- o Example: In adaptive energy grids, balance immediate load balancing with investments in renewable sources.

2. Scenario-Based Stress Testing:

- o Test systems across diverse scenarios to evaluate robustness under varying conditions.

- o Example: Simulate climate adaptation systems under extreme weather scenarios to identify vulnerabilities.

3. Dynamic Recalibration:

- o Periodically review and adjust system parameters to ensure alignment with long-term goals.

- o Example: Adaptive healthcare systems recalibrate resource allocations to prevent over-specialization during pandemics.

4. Accountability and Ethical Governance

The decentralized nature of Neodynamic systems complicates traditional accountability structures.

1. Diffuse Responsibility:

- o Distributed decision-making obscures clear accountability for failures.

- o Example: In decentralized finance, unclear governance structures make it difficult to address fraud or systemic risks.

2. Emergent Behaviors:

   o Adaptive systems may produce unintended outcomes without clear points of control.

   o Example: AI-driven trading systems inadvertently destabilize financial markets through feedback amplification.

Impact

- Undermines public trust and system legitimacy.

- Increases risks of ethical oversights or abuses.

Solutions

1. Recursive Accountability Mechanisms:

   o Embed transparent feedback loops to track and evaluate system performance.

   o Example: Urban governance systems use public dashboards to disclose decision-making processes and outcomes.

2. Independent Oversight:

   o Establish third-party organizations to monitor and evaluate adaptive systems against ethical benchmarks.

- o Example: Adaptive healthcare systems undergo periodic review by independent ethics boards.

3. Consensus Protocols:

- o Use multi-stakeholder governance models to align decentralized actions with ethical objectives.

- o Example: Adaptive climate governance integrates input from local communities, scientists, and policymakers.

Neodynamics offers a unique opportunity to embed ethics into the very fabric of adaptive systems. By addressing challenges like bias, privacy, overadaptation, and accountability, it ensures that these systems align with societal values of equity, transparency, and resilience. Ethical considerations are not limitations but integral components of Neodynamics, strengthening its ability to navigate complexity while empowering humanity.

The challenges and criticisms explored in this chapter reflect the complexity and ambition of Neodynamics. From theoretical ambiguities to implementation barriers and ethical dilemmas, these issues are not limitations but opportunities to refine the framework.

Addressing these concerns rigorously strengthens Neodynamics, ensuring its resilience and applicability across domains.

Synthesis of Challenges

Theoretical Refinement

• Ambiguities in Emergent Coherence: Ongoing research into mathematical modeling and practical metrics like the Coherence Alignment Metric (CAM) addresses these concerns, enabling precise evaluation of decentralized systems.

• Feedback Dynamics: Multi-layered feedback mechanisms mitigate risks of delays, overcorrection, or instability, ensuring systems balance adaptability with stability.

• Probabilistic Foundations: Scenario-based modeling and hybrid approaches bridge gaps between adaptability and predictive precision.

Implementation Barriers

• Technological Challenges: Modular designs, cloud computing, and synthetic data generation lower barriers for resource-constrained actors.

- Institutional Resistance: Hybrid governance models and participatory design processes foster alignment and build stakeholder trust.

- Systemic Inequalities: Redistributive mechanisms and equity-weighted metrics ensure that adaptive systems prioritize underserved communities.

Ethical Innovations

- Bias Mitigation: Continuous auditing and stakeholder inclusion ensure fairness and equity in adaptive systems.

- Privacy Protections: Privacy-by-design principles and decentralized architectures safeguard individual rights.

- Accountability Structures: Recursive accountability mechanisms and independent oversight bodies ensure transparent governance.

Opportunities for Refinement

Each critique represents an opportunity to improve Neodynamics and expand its relevance:

Cross-Disciplinary Learning

- Integrating insights from anthropology, neuroscience, and ethics deepens the theoretical underpinnings and broadens applicability.

- Example: Insights from behavioral economics inform incentive structures for decentralized systems.

Iterative Design Principles

- The principles of adaptability and feedback apply not just to systems but to Neodynamics itself, guiding its evolution through iterative testing and stakeholder feedback.

- Example: Pilot projects in urban planning or healthcare generate data that refine Neodynamic constructs.

Ethical Leadership

- Ethical considerations become central to system design, fostering trust and legitimacy. Neodynamics positions itself as a framework not just for efficiency but for societal transformation.

- Example: Transparent AI systems demonstrate how ethical safeguards enhance public trust and system adoption.

Vision for Impact

Neodynamics offers a path forward for designing systems that are not only functional but also equitable, sustainable, and aligned with human and ecological well-being. By addressing challenges as opportunities, the framework evolves into a more robust and comprehensive science of adaptability. Adaptive systems informed by Neodynamics can address pressing global challenges such as climate change, economic instability, and systemic inequities. By fostering collaboration across fields, Neodynamics provides a unifying language and methodology for tackling complexity. Ultimately, Neodynamics represents the science of human potential, enabling systems that align with the way the universe works rather than imposing rigid structures upon it.

The challenges and criticisms of Neodynamics are integral to its growth and refinement. Far from undermining its principles, they highlight the framework's relevance and inspire continuous evolution. By addressing these concerns with rigor and creativity, Neodynamics becomes not just a theory of systems but a roadmap for navigating the complexities of a dynamic, interconnected world.

Future Directions

The future of Neodynamics is as vast as the challenges and opportunities it seeks to address. Its principles of adaptability, feedback, and emergent coherence offer a universal framework for designing systems that thrive in complexity. Yet, its true potential lies in its capacity to evolve. This chapter envisions how Neodynamics can expand its theoretical constructs, validate its applications, and address pressing global challenges while exploring speculative domains that stretch the boundaries of human imagination. It is a forward-looking reflection on how this science can guide humanity through uncertainty and into a flourishing future.

Neodynamics is not merely a tool for problem-solving; it is a lens for understanding the interconnectedness of life, systems, and the cosmos. As we explore its future, we must embrace its recursive nature, where every application informs its refinement and every challenge deepens its relevance. This chapter charts a path for Neodynamics to grow, integrate, and transform, ensuring it remains a living science that adapts alongside the world it seeks to influence.

The theoretical foundations of Neodynamics are robust, but like any new science, they are far from complete. Central constructs such as the Unified Field of Adaptive Potential (UFAP) and the SPARC Framework offer a compelling starting point, but their depth and breadth can expand with continued exploration. UFAP, for example, maps adaptability in abstract, multi-dimensional terms, but future work could integrate empirical data to refine its models, creating tools that are not only conceptually elegant but also empirically grounded. Similarly, the SPARC Framework's recursive choice mechanisms could evolve into more complex decision-making models that account for interdependencies across scales and systems.

One promising direction is the formal integration of uncertainty into Neodynamic constructs. In a world increasingly defined by unpredictable events—climate disruptions, technological breakthroughs, and social upheavals—systems must not only adapt but thrive in the face of uncertainty. This requires a deeper understanding of how feedback mechanisms interact with probabilistic outcomes and how emergent coherence can persist amid chaotic inputs. Advancing these areas could position Neodynamics as a cornerstone for designing resilient systems in an increasingly volatile world.

Collaboration across disciplines will also be essential. While Neodynamics draws heavily from systems theory, thermodynamics, and information theory, its potential can only be realized through integration with other fields. Behavioral sciences, for instance, could enrich its understanding of human decision-making within adaptive systems, while advancements in computational modeling could provide new tools for testing its principles at scale. By expanding its theoretical constructs in this way, Neodynamics not only strengthens its foundations but also ensures its relevance across a growing range of applications.

For Neodynamics to achieve its full potential, it must move beyond its conceptual elegance to demonstrate tangible efficacy through validation and empirical testing. While its constructs, such as the Unified Field of Adaptive Potential (UFAP) and SPARC Framework, are theoretically robust, they must be tested in real-world scenarios and diverse domains to refine their applicability. Empirical validation ensures that Neodynamics becomes more than a theoretical discipline—it transforms into a practical science that informs decision-making and system design across sectors.

Simulation and modeling will play a critical role in this validation process. Agent-based models (ABM), system dynamics (SD), and hybrid frameworks provide controlled environments where Neodynamic principles can be tested against complex, variable conditions. These simulations not only validate constructs but also reveal emergent behaviors, potential limitations, and opportunities for refinement. For instance, simulating adaptive healthcare systems during a pandemic could demonstrate how real-time resource allocation, guided by feedback loops, outperforms traditional static approaches. Similarly, climate adaptation models could explore how Neodynamics balances ecological sustainability with economic and social equity under varying environmental pressures.

Beyond simulations, pilot programs in real-world settings are essential. These programs allow for iterative testing and refinement, providing insights into how Neodynamic systems interact with human behaviors, organizational cultures, and external conditions. For example, an adaptive governance pilot in urban planning could test decentralized decision-making frameworks, measuring their effectiveness in achieving emergent coherence across diverse neighborhoods. Similarly, adaptive supply chain initiatives could evaluate how Neodynamics responds to global disruptions, such as natural disasters or geopolitical crises, while maintaining efficiency and equity.

Interdisciplinary collaboration is another critical avenue for empirical testing. Neodynamics intersects with fields such as data science, behavioral psychology, and network theory, offering opportunities for collaborative validation efforts. By partnering with experts in these areas, researchers can design experiments that test not only the functionality of Neodynamic systems but also their impact on human outcomes, such as trust, creativity, and resilience.

Validation is not a one-time process but an ongoing, recursive effort that mirrors the principles of Neodynamics itself. Every test, simulation, and pilot program generates feedback that informs the next iteration, ensuring that the framework continues to evolve in response to new insights and challenges. This commitment to empirical testing will solidify Neodynamics as a practical and transformative science, capable of guiding systems through the uncertainties of the future.

The global challenges humanity faces today—climate change, inequality, pandemics, and technological disruption—require systems that are not just resilient but deeply adaptive and equitable. Neodynamics offers a framework for addressing these issues by embedding adaptability, feedback, and coherence into the design of human systems. Its principles are not bound by geography, culture, or discipline; they provide a universal guide for navigating complexity and fostering collaboration at every scale. This vision for global impact extends beyond immediate problem-solving to imagining a future where systems enable humanity to thrive in harmony with the planet and the cosmos.

One of the most pressing areas for Neodynamics' application is climate change. Current mitigation and adaptation strategies often rely on static models that fail to account for dynamic feedback and evolving conditions. Neodynamics introduces a paradigm shift: adaptive climate governance that responds in real time to environmental changes, resource availability, and social needs. For instance, coastal cities could employ Neodynamic frameworks to balance flood defense with ecosystem preservation, reallocating resources dynamically based on feedback from weather patterns, biodiversity metrics, and community input. By aligning human systems with natural processes, Neodynamics not only addresses the symptoms of climate change but also fosters long-term ecological harmony.

Equity is another cornerstone of Neodynamics' global vision. The framework's ability to integrate feedback-driven equity metrics ensures that systems prioritize underserved populations and regions. In healthcare, this could mean dynamically reallocating resources during a pandemic to areas with the greatest need, balancing efficiency with fairness. In education, adaptive platforms could address systemic inequities by tailoring curricula and resources to underserved schools, creating opportunities where they are most needed. These applications demonstrate how Neodynamics transforms abstract principles into actionable pathways for justice and inclusion on a global scale.

Speculative and emerging domains offer further opportunities for Neodynamics to shape the future. In quantum computing, the framework aligns with high-dimensional problem-solving, enabling real-time optimization of complex systems like global supply chains or planetary resource management. In interstellar exploration,

Neodynamic principles guide the design of self-sustaining habitats and governance structures that evolve under unknown and dynamic conditions. These speculative applications reflect the universality of Neodynamics, showing how its principles transcend current challenges to unlock new possibilities for humanity.

The true power of Neodynamics lies in its alignment with universal dynamics. By mirroring the feedback loops, adaptability, and emergent coherence found in nature and the cosmos, it offers a model for progress that is sustainable and inclusive. It reimagines global systems as collaborative and dynamic, capable of evolving alongside the complexities of an interconnected world. This vision for global impact is not just aspirational; it is achievable, provided humanity embraces the principles of Neodynamics as a guide for action and innovation.

Neodynamics is not only a framework for designing systems but also a philosophy that challenges us to rethink our relationship with complexity, uncertainty, and progress. At its core, it aligns with the natural principles of adaptability, feedback, and coherence, encouraging humanity to see itself as an integral part of the dynamic systems that govern life and the cosmos. As we expand its philosophical and ethical dimensions, Neodynamics offers profound insights into how we can foster harmony and equity while embracing the unpredictability of the future.

Philosophically, Neodynamics reframes progress. Traditional models often define progress in linear terms, focusing on material growth, control, or domination of natural systems. Neodynamics, by contrast, emphasizes alignment and balance. It teaches that true progress lies in navigating complexity with creativity and coherence, achieving dynamic stability rather than static control. This perspective draws from systems theory and natural dynamics, where adaptability is the key to survival and flourishing. By adopting this lens, humanity can redefine success—not as the elimination of uncertainty, but as the ability to thrive within it.

Ethically, Neodynamics demands a commitment to equity and inclusion. Feedback loops that prioritize fairness ensure that systems address historical inequities rather than perpetuate them. For instance, adaptive governance frameworks could dynamically redistribute resources to underserved communities, integrating ethical considerations into their very structure. Similarly, in artificial intelligence, Neodynamics provides a roadmap for embedding ethical safeguards into adaptive decision-making processes, ensuring systems remain accountable and aligned with human values. These examples highlight how Neodynamics not only responds to ethical challenges but also embeds moral responsibility into the fabric of its systems.

One of the most transformative ethical principles of Neodynamics is its call for stewardship. By aligning with the adaptive logic of natural and cosmic systems, it positions humanity not as a dominant force but as a collaborative participant in the broader web of life. This philosophy echoes Indigenous perspectives that emphasize living in harmony with nature, rather than exploiting it. It also challenges modern societies to move beyond extractive practices and embrace regenerative approaches to resource management, governance, and innovation.

Neodynamics also expands the ethical conversation around speculative domains. As humanity ventures into uncharted territories—such as space exploration or quantum computing—it faces unprecedented moral dilemmas. How should adaptive systems balance autonomy and coherence in interstellar colonies? What safeguards are needed to ensure quantum systems respect human agency and fairness? Neodynamics provides a foundation for addressing these questions, framing ethics not as an afterthought but as a central consideration in the design and evolution of complex systems.

In expanding its philosophical and ethical horizons, Neodynamics invites humanity to reimagine its role in the universe. It asks us to embrace complexity, to act as stewards of dynamic systems, and to design with equity and harmony in mind. By doing so, we align ourselves with the principles that sustain life and creativity, ensuring that progress is both meaningful and sustainable.

The future of Neodynamics is boundless, shaped by its capacity to adapt, evolve, and refine itself. As a science of potential, it offers a framework not only for solving immediate challenges but also for navigating the uncertainties and opportunities of tomorrow. Its principles of adaptability, feedback, and coherence are universal, capable of transforming systems across domains, scales, and disciplines. Yet, the journey of Neodynamics is far from complete; it is an evolving science, one that grows through the very processes it describes.

What makes Neodynamics uniquely suited for the future is its recursive nature. Every application, every experiment, and every challenge it encounters serves as a feedback loop, refining its constructs and expanding its reach. This iterative growth ensures that Neodynamics remains relevant in a world defined by rapid change and increasing complexity. It also positions the framework as a living science, one that evolves alongside the systems it seeks to influence.

The promise of Neodynamics lies in its ability to align humanity with the forces of nature and the cosmos. It teaches us to see uncertainty not as a threat but as a source of potential, to embrace complexity as an opportunity for innovation, and to design systems that thrive in harmony with their environments. From adaptive governance to interstellar exploration, from equitable healthcare to quantum computing, Neodynamics offers pathways for creating systems that empower and sustain.

As we look to the future, the task is clear: to test, refine, and expand the principles of Neodynamics. This requires collaboration across disciplines, cultures, and perspectives, as well as a commitment to ethical and inclusive practices. It also requires courage—the courage to rethink traditional models, to embrace uncertainty, and to envision progress not as domination but as flourishing.

Neodynamics is a science for an interconnected world, a guide for thriving in complexity, and a call to action for designing systems that align with the principles of life itself. Its future is in our hands, and its potential is infinite. Let this be the beginning of a journey—a journey of creation, adaptation, and coherence, guided by the dynamic forces that shape our universe and our place within it.

Conclusion

Neodynamics, as presented in this text, is both a framework and a vision for navigating the complexities of our interconnected world. It unites adaptability, feedback, and coherence into a science that transcends disciplines, challenges traditional paradigms, and offers a roadmap for dynamic systems across all domains of life.

However, it is impossible to encompass the full breadth of its applicability in a single volume. The universality of Neodynamics ensures that its principles resonate far beyond the examples and theories explored here. From the micro-scale interactions of biological systems to the macro-scale governance of interstellar colonies, the scope of its impact is as vast as it is profound. While this book strives to provide a foundational understanding, it is only the beginning. The potential for expansion and discovery is limited only by the imagination and ingenuity of those who engage with its principles. This chapter synthesizes the key contributions of Neodynamics, reflects on its human and universal significance, and offers a call to action for future exploration.

Neodynamics is a profound unification of principles and practices designed to navigate the complexities of dynamic systems.

Its core constructs, adaptability, feedback-driven processes, and emergent coherence, provide a framework for understanding how systems can thrive in environments defined by uncertainty and change. By synthesizing ideas from systems theory, thermodynamics, and information theory, Neodynamics establishes a robust theoretical foundation that transcends traditional disciplinary boundaries.

At the heart of Neodynamics lies the Unified Field of Adaptive Potential (UFAP), a conceptual model mapping the infinite pathways systems can take in response to external pressures and internal objectives. UFAP transforms adaptability from an abstract idea into a structured, navigable space, allowing systems to balance exploration and exploitation in their decision-making. Complementing UFAP is the SPARC Framework, which operationalizes adaptability through a dual focus on the Spectrum of Possibility and Recursive Choice. Together, these constructs empower systems to self-correct, recalibrate, and evolve without losing coherence.

Practical applications of Neodynamics underscore its transformative potential. Adaptive governance systems, for instance, demonstrate how feedback mechanisms can align local autonomy with global objectives, ensuring resilience and responsiveness in the face of crises. Healthcare systems leveraging Neodynamics dynamically allocate resources during pandemics, balancing efficiency with equity to protect vulnerable populations. Similarly, in education, adaptive learning platforms respond to individual progress, fostering creativity and inclusivity while bridging systemic inequalities. These examples illustrate how Neodynamics can address some of the most pressing challenges of our time by embedding adaptability and coherence into the fabric of human systems.

Beyond immediate applications, Neodynamics extends into speculative and emerging domains, pushing the boundaries of what is possible. Quantum computing offers an avenue for modeling adaptive systems at an unprecedented scale, enabling real-time feedback and optimization in complex networks. Interstellar exploration relies on Neodynamic principles to design self-sustaining habitats and governance frameworks capable of evolving under unknown conditions. In post-capitalist economic systems, dynamic redistribution mechanisms align resources with need, fostering equity and sustainability. These speculative applications not only showcase the versatility of Neodynamics but also invite future research and experimentation.

The principles of Neodynamics echo patterns found in nature, linking it to universal dynamics. Feedback loops drive evolution, emergent coherence governs ecosystems, and adaptability ensures survival across scales—from cellular processes to planetary systems. By mirroring these natural processes, Neodynamics aligns human systems with the broader forces of life and the cosmos, offering a roadmap for sustainable progress.

In synthesizing its contributions, Neodynamics emerges as a science of potential. It bridges theory and practice, addressing immediate challenges while opening new frontiers for exploration. By integrating adaptability, feedback, and coherence, Neodynamics not only transforms how systems are designed but also redefines how humanity understands and interacts with complexity. Its contributions are as much a call to action as they are a foundation for future discovery.

Neodynamics is more than a framework for designing adaptive systems; it is a philosophy that resonates deeply with human aspirations. At its core, it reflects the fundamental human need for agency, equity, creativity, and resilience. In a world often defined by complexity and uncertainty, Neodynamics offers a way to not only navigate these challenges but also to thrive within them. By prioritizing adaptability and coherence, it provides tools for designing systems that respond dynamically to human needs and foster a sense of belonging and empowerment.

The principle of human agency is central to Neodynamics. Adaptive systems that integrate feedback loops restore control to individuals and communities, enabling them to shape their environments rather than be constrained by them. In education, for example, adaptive learning platforms empower students to take ownership of their growth, adjusting in real time to individual strengths and challenges. In healthcare, systems that dynamically recalibrate based on patient feedback prioritize personalized care, enhancing trust and satisfaction. These systems embody the Neodynamic ethos: designing with, not just for, the people they serve.

Equity and inclusivity are essential dimensions of Neodynamics, ensuring that adaptive systems do not perpetuate or exacerbate existing disparities. By embedding equity metrics into feedback mechanisms, Neodynamics prioritizes underserved populations, reallocating resources dynamically to where they are needed most. In urban governance, adaptive housing systems address homelessness by responding to real-time data on shelter needs, while in healthcare, equitable resource distribution ensures marginalized communities receive timely and fair access to care. These examples demonstrate how Neodynamics can create systems that are not only efficient but also just, fostering hope and societal cohesion.

Creativity and innovation flourish in adaptive environments, and Neodynamics provides the conditions for this growth. By embracing feedback and iterative refinement, systems designed with Neodynamics promote experimentation and the exploration of new possibilities. In generative AI, adaptive frameworks ensure outputs remain dynamic, ethical, and culturally relevant, empowering creators to push boundaries. In cultural systems, decentralized feedback-driven platforms amplify diverse voices, fostering vibrant and resilient communities. These systems inspire curiosity and ingenuity, reflecting the human desire to grow, create, and connect.

Resilience is another key promise of Neodynamics. In the face of crises, adaptive systems provide stability and a sense of security. Coastal cities employing adaptive flood defenses, for example, protect vulnerable populations while preserving ecological integrity, demonstrating the power of dynamic resource management. Adaptive healthcare systems reduce systemic strain during pandemics by reallocating resources in real time, ensuring preparedness and reducing public anxiety. These systems embody the capacity of Neodynamics to foster resilience and hope, even in the most challenging circumstances.

Ultimately, Neodynamics offers a vision of collective flourishing. It moves beyond survival to design systems that enable individuals and communities to thrive. Adaptive education systems cultivate lifelong learning, while equitable healthcare systems ensure well-being for all. These principles, applied globally, align humanity with the dynamic forces of life, creating a world where progress is not measured by material growth alone but by creativity, equity, and harmony. Neodynamics is a science of human potential, offering a path toward a flourishing future for all.

Neodynamics transcends the boundaries of human systems and touches upon the fundamental principles that govern the universe itself. It is a framework that mirrors the adaptability, feedback, and emergent coherence inherent in natural and cosmic processes. By aligning human endeavors with these universal dynamics, Neodynamics offers not only a guide for thriving within complexity but also a philosophy for understanding our place in the broader fabric of existence.

The natural world provides countless examples of the principles underlying Neodynamics. In ecosystems, feedback loops regulate balance, ensuring that species adapt and evolve in harmony with their environments. Coral reefs, for instance, adjust to rising temperatures by reallocating energy within their systems, demonstrating the interplay between adaptation and coherence. Similarly, the water cycle, with its constant feedback and redistribution, exemplifies how natural systems self-regulate to sustain life. These patterns, observed in nature, are the same principles that Neodynamics applies to human systems, bridging the micro-level dynamics of biological processes with the macro-scale challenges of governance and technology.

At a larger scale, the universe itself reflects the principles of Neodynamics. The formation of galaxies, the adaptive expansion of the cosmos, and the gravitational feedback loops that shape planetary systems all embody emergent coherence and adaptability. These phenomena suggest that the forces governing the universe are inherently dynamic, responding to shifts in energy and matter to maintain balance and foster complexity. By understanding and applying these principles, humanity can design systems that are not only efficient but also in harmony with the fundamental order of the cosmos.

Neodynamics also provides a philosophical framework for progress. Traditional concepts of progress often focus on linear growth or control, but Neodynamics reframes progress as a dynamic balance between exploration and coherence. This balance is evident in natural systems, where stability and change coexist to foster resilience. Adaptive human systems, such as urban planning that integrates green spaces and renewable energy, mirror these dynamics, aligning human development with ecological principles. Neodynamics teaches that progress is not about mastering complexity but about thriving within it, leveraging adaptability to sustain harmony.

Human systems are not separate from universal dynamics; they are microcosms of the same principles that govern the cosmos. By adopting Neodynamics, humanity can align its actions with the natural order, creating systems that are not only sustainable but also reflective of the universe's inherent adaptability. For instance, decentralized governance models echo the self-organizing patterns found in ecosystems, demonstrating how local autonomy and global coherence can coexist.

Ultimately, Neodynamics positions humanity as stewards of complexity, capable of creating systems that sustain life and harmony across scales. It offers a roadmap for bridging the human, the natural, and the cosmic, fostering a deeper understanding of our interconnectedness. By aligning with the principles of adaptability, feedback, and coherence, Neodynamics empowers humanity to thrive as part of the universal dynamic rather than in opposition to it.

As transformative as Neodynamics is, its implementation is not without challenges. These hurdles, however, are not insurmountable; they present opportunities for refinement, growth, and innovation. This section outlines the theoretical, practical, and ethical challenges facing Neodynamics while framing them as steps in its evolution from a conceptual framework to a universal science of adaptation and coherence.

Theoretical challenges lie at the heart of advancing Neodynamics. While constructs like the Unified Field of Adaptive Potential (UFAP) and SPARC Framework are conceptually robust, their empirical validation remains a critical next step. Testing these constructs through simulation models and real-world pilot programs will uncover nuances that theoretical explorations cannot predict. For instance, modeling adaptive resource allocation during crises, such as pandemics or natural disasters, provides an invaluable testing ground for refining these principles. Similarly, ensuring that emergent coherence sustains itself under extreme conditions—such as in systems facing prolonged stress or conflicting objectives— requires rigorous stress testing and scenario modeling. These theoretical gaps are not limitations but opportunities for interdisciplinary collaboration, inviting researchers to integrate insights from mathematics, computer science, and behavioral studies into Neodynamics' growth.

Practical implementation presents another layer of challenges. Resistance to change, particularly in entrenched institutions, can hinder the adoption of adaptive systems. Organizations and governments accustomed to static hierarchies may view Neodynamic systems, with their decentralized decision-making and iterative recalibrations, as disruptive. Demonstrating small-scale successes can alleviate these concerns, as seen in adaptive energy grids that begin at the neighborhood level and gradually expand to entire cities. Similarly, the scalability of Neodynamic systems—how localized adaptations integrate into global objectives—remains an area for exploration. Multi-scale modeling frameworks that balance local autonomy with global coherence offer a pathway forward, ensuring that adaptability at one scale does not conflict with broader systemic goals.

Ethical considerations are integral to the success of Neodynamics. Adaptive systems must be designed to avoid reinforcing existing biases or inequities. If feedback mechanisms are poorly structured or data inputs are biased, these systems risk perpetuating the very issues they aim to solve. Embedding equity metrics as foundational parameters can address this challenge, dynamically reallocating resources to underserved populations or regions. Another ethical concern is accountability in decentralized systems. When decision-making is distributed, determining responsibility for failures or ethical breaches becomes complex. Transparent feedback loops, coupled with governance protocols that emphasize accountability, can mitigate these risks.

Despite these challenges, the opportunities for growth are immense. Global crises—such as climate change, pandemics, and economic instability—are not only tests of existing systems but also proving grounds for Neodynamics. Adaptive disaster response frameworks, for instance, demonstrate the efficacy of feedback-driven systems in reallocating resources dynamically during emergencies. Emerging technologies, such as quantum computing and AI, further amplify Neodynamics' potential, enabling real-time processing of complex feedback loops at scales previously unimaginable. Interdisciplinary research initiatives provide another avenue for growth, fostering collaboration between fields and unlocking new applications for adaptive systems.

Challenges are not obstacles but essential steps in the evolution of Neodynamics. Each hurdle—whether theoretical, practical, or ethical—pushes the framework to adapt and grow, embodying its own principles of feedback and coherence. By addressing these challenges, Neodynamics evolves from a theoretical construct into a transformative force, ready to reshape how humanity designs systems and navigates complexity.

Neodynamics is more than a framework for solving today's challenges; it is a blueprint for navigating the complexities of tomorrow. Its principles of adaptability, feedback, and emergent coherence provide a foundation for designing systems that are not only resilient but also capable of thriving in dynamic environments. This vision extends beyond immediate applications to imagine a future where humanity aligns with the fundamental forces of the universe, fostering sustainable progress and collective flourishing.

The future of Neodynamics lies in its capacity to evolve. As more systems adopt its principles, iterative refinement through empirical validation and interdisciplinary collaboration will strengthen its constructs. The Unified Field of Adaptive Potential (UFAP), for example, will gain greater precision as it is tested in diverse domains, from climate adaptation to quantum computing. Similarly, the SPARC Framework's recursive choice mechanisms will refine decision-making processes in systems ranging from healthcare to global governance. This ongoing evolution ensures that Neodynamics remains relevant, flexible, and effective in addressing the challenges of an ever-changing world.

In envisioning the future, Neodynamics redefines progress as a balance between exploration and coherence. Traditional systems often prioritize short-term efficiency or growth at the expense of long-term stability and equity. Neodynamics offers an alternative: a model where systems evolve dynamically to sustain both individual and collective well-being. This vision is already evident in adaptive education platforms that cultivate lifelong learning, in decentralized governance systems that empower local communities, and in equitable healthcare networks that prioritize underserved populations. These examples illustrate how Neodynamics can create a future defined by inclusivity, creativity, and harmony.

The speculative potential of Neodynamics pushes the boundaries of what humanity can achieve. In space exploration, adaptive systems guided by Neodynamic principles will enable sustainable settlements on Mars and beyond, ensuring resource equity and emergent social coherence. In quantum computing, real-time feedback mechanisms will optimize complex systems, from global logistics to interstellar navigation. These speculative applications reflect the transformative power of Neodynamics to unlock possibilities that transcend current limitations, bridging the known and the unknown.

At its core, Neodynamics embodies a universal philosophy of alignment. It teaches us to live in harmony with complexity, embracing uncertainty as a source of potential rather than a threat. By mirroring the adaptability and coherence inherent in nature and the cosmos, Neodynamics offers a path for humanity to act as stewards of life and complexity. This alignment is not only technical but also ethical, ensuring that systems designed with Neodynamics prioritize equity, sustainability, and collective flourishing.

The vision for the future of Neodynamics is one of empowerment and harmony. It is a call to action for researchers, practitioners, and policymakers to embrace its principles and expand its applications. By doing so, humanity can transcend the limitations of static systems and unlock its potential to thrive in complexity. Neodynamics is not just a framework for adaptation; it is a philosophy for flourishing, offering a guide for humanity to align with the dynamic forces that shape our world and the universe beyond.

As I bring this book to a close, I reflect on the journey that has unfolded within these pages—a journey into the heart of complexity, adaptability, and coherence. Neodynamics is not just a framework for designing systems; it is a science of potential and agency. It is a way of thinking and acting that aligns humanity with the dynamic principles that govern life, nature, and the cosmos. Writing this has been an act of belief—belief in the idea that we are not powerless in the face of complexity but are, instead, uniquely equipped to thrive within it.

I know this work is far from exhaustive. The scope of Neodynamics is too vast, its principles too universal, for any single book to encompass. There are theories unexplored, applications untested, and questions unanswered. Yet, this incompleteness does not diminish its value; it reinforces its potential. The universality of Neodynamics means it is not fixed or static but a living framework that will grow, evolve, and refine itself through the contributions of those who engage with it.

Neodynamics is, at its core, an acknowledgment of our place in a dynamic and interconnected world. It rejects the illusion of control and instead embraces the power of alignment—alignment with natural processes, with each other, and with the forces that shape the universe itself. It is a philosophy that teaches us to work with complexity, to see uncertainty as an invitation to innovate, and to recognize that adaptability is not a compromise but a strength.

I see Neodynamics as a science of humanity because it is a science of flourishing. Its principles are not merely technical; they are deeply human. They empower us to create systems that respond to our needs, amplify our creativity, and honor our shared values. Whether it is in designing adaptive healthcare, fostering equitable governance, or imagining self-sustaining interstellar colonies, Neodynamics invites us to rethink what is possible—not as individuals but as a collective, united by shared challenges and boundless potential.

In writing this book, I have tried to embody the principles of Neodynamics itself. Every section has been an iteration, every chapter an adaptation, and the structure as a whole an attempt to achieve coherence in the face of a vast, interconnected subject. This book, like the systems it describes, is recursive. It builds on feedback, evolves through exploration, and strives for coherence. It is not an endpoint but a beginning—a node in a larger network of ideas that will grow and change through engagement and action.

The future is not static. It is not predetermined. It is a space of infinite potential—a Unified Field of Adaptive Potential, waiting to be explored. Neodynamics is our guide to navigating that space, to embracing the feedback of our actions, and to achieving coherence in the face of complexity.

This is the science of potential. This is the science of flourishing. This is Neodynamics. Let us create, adapt, and thrive. Together.

Appendix A: Mathematical Models and Frameworks

1. The Unified Field of Adaptive Potential (UFAP)

The Unified Field of Adaptive Potential (UFAP) models the adaptability of a system as a high-dimensional space, where each dimension represents a variable that influences the system's capacity to adapt. Mathematically, UFAP can be represented as:

$$P = \{(x_1, x_2, ..., x_n) \mid x_i \in [a_i, b_i]\}$$

Where:

- P is the potential space of adaptability.

- x_i represents a variable (e.g., resource availability, environmental constraints).

- $[a_i, b_i]$ defines the range of possible states for x_i.

Optimization Within UFAP: To identify optimal adaptive strategies, UFAP employs optimization techniques. For example, a system may maximize adaptability (A) given constraints (C):

$$max A(x) \quad subject\ to\ C(x) \leq k$$

Here:

- $A(x)$ measures adaptability as a function of system variables.

- $C(x)$ represents constraints (e.g., resource limits, time).

- k is the threshold of acceptable constraint values.

Example: In an adaptive healthcare system:

- x_1: Resource availability (e.g., ventilators).

- x_2: Demand variability (e.g., patient influx).

- Optimization maximizes resource allocation efficiency while maintaining equity.

2. The SPARC Framework

The SPARC Framework operationalizes adaptability through recursive decision-making. It combines two components:

- **Spectrum of Possibility** (S): All potential responses or outcomes.

- **Recursive Choice** (R): Feedback-driven iterative refinement.

Recursive Algorithm for Decision-Making: Given a decision space (D) and feedback (F), SPARC iteratively refines choices:

1. Initialize:

$$R_0 = \frac{argmaxU(d)}{d \in D}$$

   o Where $U(d)$ is the utility function of decision d.

2. Update with Feedback:

$$R_{t+1} = R_t + \alpha F(R_t)$$

- α: Learning rate.

- $F(R_t)$: Feedback function measuring the deviation of R_t from system objectives.

3. Convergence:

$$\frac{lim}{t \to \infty} R_t = R^*$$

- R^*: Stable choice aligned with system goals.

Example: In adaptive education:

- D: Set of learning strategies.

- F: Feedback on student performance.

- SPARC refines strategies iteratively based on outcomes.

3. Feedback Loops and Stability Metrics

Adaptive systems rely on feedback to maintain stability. Feedback can be modeled using differential equations:

$$\frac{dx}{dt} = f(x, F(x))$$

Where:

- x: State of the system.

- $F(x)$: Feedback function.

- f: Dynamic adjustment based on feedback.

Stability Analysis: Stability is determined by evaluating the eigenvalues of the Jacobian matrix (J) of the system:

$$J = \frac{\partial f}{\partial x}$$

- If all eigenvalues have negative real parts, the system is stable.

Example: In urban governance:

- x: Traffic flow.

- $F(x)$: Congestion feedback.

- Stability analysis ensures adaptive traffic systems avoid oscillations or chaotic behavior.

4. Emergent Coherence in Multi-Agent Systems

Emergent coherence arises from decentralized interactions between agents. It can be modeled using agent-based approaches:

Interaction Rule:

$$x_i(t + 1) = x_i(t) + \sum_j w_{ij}(x_j(t) - x_i(t))$$

Where:

- $x_i(t)$: State of agent i at time t.

- w_{ij}: Weight of influence from agent j to i.

Emergent Metric: Coherence is measured using an order parameter (O):

$$O = \frac{1}{N}\left|\sum_{i=1}^{N} e^{i\theta i}\right|$$

- $O = 1$: Perfect coherence.

- $O = 0$: Complete disorder.

Example: In decentralized energy grids:

- Agents represent local nodes.

- x_i: Energy demand at node i.

- Interaction rules balance supply and demand dynamically.

Appendix B: Simulation and Modeling Techniques

1. Agent-Based Modeling (ABM)

Overview:

Agent-based modeling simulates interactions among autonomous agents within a defined environment, allowing for the emergence of complex, system-wide behaviors. Each agent operates under a set of rules, adapting to changes in its environment or other agents.

Key Steps:

1. **Define Agents**: Specify agent types and attributes.

   o **Example:** In adaptive healthcare, agents might represent hospitals, patients, and supply chains.

2. **Set Interaction Rules**: Define how agents interact with each other and their environment.

   o **Example:** Patients interact with hospitals based on proximity and resource availability.

3. **Simulate Dynamics**: Observe emergent behaviors over time.

   o **Example:** Resource allocation shifts dynamically as patient demand changes.

4. **Analyze Outcomes**: Evaluate metrics such as system efficiency or equity.

Tools:

- **NetLogo**: Easy-to-use ABM software for conceptual modeling.

- **Repast**: Advanced platform for large-scale, computationally intensive ABM.

- **AnyLogic**: Combines ABM with other methodologies like system dynamics.

2. System Dynamics (SD)

Overview:

System dynamics focuses on the behavior of interconnected variables over time, using differential equations to model feedback loops and delays.

Key Steps:

1. **Identify Variables**: Determine stocks (e.g., resources) and flows (e.g., resource consumption rates).

o **Example:** In urban governance, stocks might include housing availability, and flows could represent construction or migration.

2. **Establish Relationships**: Map causal links between variables.

   o **Example:** Higher housing costs might reduce migration rates.

3. **Incorporate Feedback**: Include reinforcing (positive) and balancing (negative) feedback loops.

   o **Example:** More housing reduces demand pressure, stabilizing prices (balancing feedback).

4. **Simulate Over Time**: Observe trends and stability.

   o **Example:** Evaluate how different housing policies impact urban growth.

Tools:

- **Vensim**: A user-friendly platform for creating and analyzing SD models.

- **Stella**: Focuses on intuitive modeling for education and policymaking.

3. Hybrid Approaches

Overview:

Hybrid modeling integrates ABM and SD to combine micro-level agent behaviors with macro-level system dynamics. This approach captures both local interactions and global trends.

Example:

- **Healthcare Resource Allocation**:

    o **ABM**: Simulates patient movement between hospitals.

    o **SD**: Tracks overall resource trends, such as medical supply levels.

    o **Outcome**: Identifies bottlenecks and ensures equitable distribution.

Benefits:

- Captures emergent behaviors from agent interactions (ABM).
- Models long-term trends and systemic dynamics (SD).
- Useful for complex systems like adaptive energy grids or disaster response frameworks.

Tools:

- **AnyLogic**: Built for hybrid modeling, seamlessly integrating ABM and SD components.

4. Real-Time Simulations and Predictive Analytics

Overview:

Real-time simulations integrate live data streams to adapt dynamically during runtime, enabling systems to respond to ongoing changes.

Applications:

1. **Disaster Response**:

   o **Integrates weather, population density, and resource availability data to direct aid dynamically.**

2. **Energy Grids**:

   o **Uses live consumption data to balance supply and demand adaptively.**

Predictive Analytics:

- Uses historical data and machine learning to forecast system behavior and guide decision-making.

- Example: Predictive models in adaptive agriculture anticipate droughts and optimize irrigation schedules accordingly.

Tools:

- **Simulink**: Used for real-time simulations in engineering systems.

- **Python (Pandas, TensorFlow)**: Combines simulation with predictive modeling for data-driven adaptability.

5. Visualization and Analysis

Overview:

Effective visualization tools enhance understanding of simulation results, allowing stakeholders to interpret system behaviors and outcomes intuitively.

Tools:

1. **Gephi**: For network visualizations of agent interactions.

2. **Matplotlib (Python)**: For creating time-series and causal loop diagrams.

3. **Tableau**: For interactive dashboards showcasing real-time data and trends.

6. Ethical Considerations in Simulation

Transparency:

Simulations must clearly document assumptions, data sources, and limitations to avoid misleading interpretations.

Equity:

Ensure simulations do not reinforce biases, particularly in adaptive systems involving sensitive areas like healthcare or education.

Accountability:

Stakeholders must validate simulation outcomes, especially when used in policymaking or resource allocation.

Conclusion

Simulation and modeling are indispensable tools for testing and refining Neodynamic systems. By integrating methodologies like ABM, SD, and real-time analytics, these techniques provide insights into the complexities of adaptive systems and their emergent behaviors.

Appendix C: Glossary of Key Terms

Adaptability

The ability of a system to adjust its structure, behavior, or parameters in response to changing conditions, enabling it to maintain functionality or thrive in dynamic environments.

Agent-Based Modeling (ABM)

A simulation method that models the interactions of autonomous agents, capturing emergent behaviors within complex systems.

Balancing Feedback Loop

A type of feedback loop that stabilizes a system by counteracting changes, maintaining equilibrium.

Coherence (Emergent Coherence)

The harmonious alignment of a system's components, achieved through decentralized interactions and feedback-driven processes, resulting in functional order.

Decentralized Governance

A governance model where decision-making authority is distributed among multiple entities or local units, promoting flexibility and responsiveness.

Dynamic Stability

A system's capacity to maintain functional coherence while adapting to changing conditions, balancing resilience and flexibility.

Emergent Behavior

Complex system-wide patterns that arise from the interactions of individual components or agents without centralized control.

Equity Metrics

Quantitative measures that assess the fairness and inclusivity of a system's resource distribution, decision-making, or outcomes.

Feedback

Information about a system's current state that is used to adjust its future behavior. Feedback can be positive (amplifying change) or negative (stabilizing).

Feedback-Driven Processes

Mechanisms that rely on real-time feedback to guide a system's decisions and behaviors, enabling continuous adaptation and improvement.

Hybrid Modeling

A simulation approach that integrates multiple methodologies, such as agent-based modeling and system dynamics, to capture both micro-level interactions and macro-level trends.

Recursive Choice

An iterative process in which a system refines its decisions based on feedback, progressively aligning its actions with changing objectives.

Self-Regulating System

A system that autonomously adjusts its behavior through internal mechanisms, such as feedback loops, to achieve stability or adapt to external changes.

Spectrum of Possibility

The range of potential outcomes or responses a system can generate, forming the foundation for adaptability and decision-making.

SPARC Framework

A practical implementation of Neodynamics that combines the Spectrum of Possibility and Recursive Choice to create adaptive systems.

Stability Metrics

Quantitative measures used to evaluate the robustness and resilience of a system, often based on feedback dynamics and equilibrium states.

System Dynamics (SD)

A simulation methodology that models the behavior of interconnected variables over time, focusing on feedback loops and delays.

Unified Field of Adaptive Potential (UFAP)

A conceptual framework that represents the multidimensional space of all possible adaptive actions a system can take in response to dynamic conditions.

Universal Alignment

A philosophical principle emphasizing the alignment of human systems with the fundamental dynamics of nature and the cosmos, such as adaptability and coherence.

Urban Adaptation

The application of adaptive systems in urban planning and governance, enabling cities to respond dynamically to challenges such as population growth, resource constraints, and environmental pressures.

Vision for Flourishing

A central tenet of Neodynamics that focuses on enabling systems to thrive beyond mere survival, fostering creativity, equity, and harmony.

Weight of Influence

In agent-based models, a parameter that quantifies the degree to which one agent's behavior affects another, influencing emergent dynamics.

References

Books and Foundational Texts

Ashby, W. R. (1956). *An introduction to cybernetics*. Chapman & Hall.

Meadows, D. H. (2008). *Thinking in systems: A primer*. Chelsea Green Publishing.

Maturana, H. R., & Varela, F. J. (1972). *Autopoiesis and cognition: The realization of the living*. D. Reidel Publishing Company.

Shannon, C. E. (1948). *A mathematical theory of communication*. The Bell System Technical Journal, 27, 379–423, 623–656.

Sterman, J. D. (2000). *Business dynamics: Systems thinking and modeling for a complex world*. McGraw Hill.

Wiener, N. (1948). *Cybernetics: Or control and communication in the animal and the machine*. MIT Press.

Wiener, N. (1954). *The human use of human beings: Cybernetics and society*. Houghton Mifflin Harcourt.

Articles and Journals

Bar-Yam, Y. (1997). *Dynamics of complex systems*. Perseus Books.

Forrester, J. W. (1961). *Industrial dynamics*. MIT Press.

Holland, J. H. (1992). Complex adaptive systems. *Daedalus*, 121(1), 17–30.

Simon, H. A. (1962). The architecture of complexity. *Proceedings of the American Philosophical Society*, 106(6), 467–482.

Journals and Conference Proceedings

System Dynamics Society. (n.d.). *System dynamics review.* Retrieved from https://onlinelibrary.wiley.com/journal/10991727

Santa Fe Institute. (n.d.). Complexity journal. Retrieved from https://www.santafe.edu/research/results/complexity-journal

Journal of Artificial Societies and Social Simulation (JASSS). (n.d.). Retrieved from https://www.jasss.org

Online Resources and Platforms

Center for Adaptive Systems Applications (CASA). (n.d.). Adaptive systems in action. Retrieved from https://www.casatechnology.com

MIT OpenCourseWare. (n.d.). Dynamics of complex systems by John Sterman. Retrieved from https://ocw.mit.edu

NetLogo Modeling Platform. (n.d.). Retrieved from https://ccl.northwestern.edu/netlogo/

Python Mesa Framework. (n.d.). Agent-based modeling in Python.

Retrieved from https://mesa.readthedocs.io

www.ingramcontent.com/pod-product-compliance
Lightning Source LLC
Chambersburg PA
CBHW071447220526
45472CB00003B/704